Laboratory Manual for Electrical Machines

Laboratory Manual for Electrical Machines

Second Edition

D.P. Kothari
FNAE, FNASc, Fellow-IEEE, Hon. Fellow, ISTE, FIE (India)
Dean (Research and Development)
J D College of Engineering and Management, Nagpur
Ex Director - Research
S B Jain Institute of Management and Research, Nagpur
Former Director Research, MVSR Engg. College, Hyderabad
Former Director General
J B Group of Educational Institutions, Hyderabad
Former DG, RGI, Nagpur
Emeritus Director General, VITS Indore
Former VC, VIT Vellore
Former Director i/c IIT Delhi
Former Principal VRCE, Nagpur

B.S. Umre
Associate Professor
Department of Electrical Engineering, VNIT
Nagpur

TECHSAR™
NEW DELHI

Published and Printed by
Techsar Pvt. Ltd.
4435-36/7, Ansari Road, Daryaganj,
New Delhi- 110 002 (India)
E-mail: tech_info@techsarworld.com
Website: www.techsarworld.com

ISBN 978-93-85909-75-7

Reprint 2025

Dedicated to
My In-laws
Late Shri R. M. Somani
Late Smt. Kusum Somani

D.P. Kothari

My Parents
Smt. Savitribai Sitaram Umre
Late Shri Sitaram Pandurang Umre

B.S. Umre

Preface to the Second Edition

The excellent response to the first edition of the book by students and faculty of Indian and foreign universities and the practising engineers has motivated us to venture for the second edition. The main aim was to include some more experiments in electrical machines so that it can cater to the complete syllabus of undergraduate laboratory course of electrical machines.

As compared to the first edition (2013) in this edition four new experiments are added. These are basic experiments and are useful for new entrants in engineering. The basic laws of electrical engineering are KCL and KVL. The B-H characteristic is the basic of all electrical machines. The main components or elements of electrical engineering are resistor, inductor and capacitor. The responses of these components under sinusoidal excitation are explained in details. The understanding of phasor and waveform methods for these components are very useful for solving the complex problems in electrical engineering.

In short, these new four experiments are useful to understand the basics of electrical engineering.

We are grateful to the authorities of JD College of Engineering and Management Nagpur (Shri Sanjay Agrawal, Chairman) and the authorities of VNIT Nagpur (Prof. Dr. N. S. Chaudhari, the Director) for giving all the help and encouragement for writing this book.

D. P. Kothari
B. S. Umre

Preface to the First Edition

Electrical machines form an integral part of a large electrical power system, with other components as well. The aim of this book is to give detailed theory of the experiments on electrical machines. As a step in this direction, we have great pleasure in publishing this laboratory manual for electrical engineering students of B. E. and B. Tech. and polytechnics in India and abroad. This will give the basic information to the students with the machine phenomena, working principles and testing methods, etc. This will impart a real physical understanding of various types of electrical machines.

The main attraction of this laboratory manual is its Power Point Presentation for all experiments. With the help of this PPT, any college or institute can conduct these experiments using available machines with minor modifications. By studying this PPT, an average student can also perform the experiments with great interest and enthusiasm.

The experiments in this manual are broadly divided as:

1. Experiments in the second year of B.E. or B.Tech.
2. Experiments in the third year of B.E. or B.Tech.
3. Second and the third year of polytechnics.

The conclusion written at the end will help the student as well as working engineers to know the behaviour of various machines. We thank our families who supported us during this period and gave all possible help so that this book can see the light of the day.

The authors are thankful to the employers for their enocouragement and help in bringing out this book.

We welcome constructive criticism of the book and will be grateful for an appraisal by the readers.

D. P. Kothari
B. S. Umre

Contents

General Precautions

General precautions to be observed while doing the experiments:

1. Understand the **equipment** to be tested and **apparatus** to be used.
2. Select proper type (i.e., AC or DC) and range of meters.
3. **Do not touch live wires and terminals.**
4. Use suitable wires (type and size), e.g., thin flexible wires for voltmeters.
5. All the connections should be **tight.**
6. Do not leave loose wires (i.e., wires not connected).
7. Get the connection **checked** by the teacher before switching 'ON' the supply.
8. **Never exceed** the permissible values of current, voltage and / or speed of any machines meters, apparatus, wires, loads, etc.
9. **Switch on or switch off the load gradually and not suddenly.**
10. Do not touch the meters while noting down the readings.
11. **Switch off** the supply after completion of the experiment.

CHAPTER 1

To Study the Speed Control of a DC Shunt Motor by

(a) Varying armature voltage with field current kept constant
(b) Varying field current with armature voltage kept constant

1. OBJECTIVE

The main objective of this experiment is to see the effects of field current I_f and armature voltage V_a on the speed of DC shunt motor.

2. PRE-EXPERIMENTAL QUIZ

1. Write down the equation of speed for DC shunt motor.
2. Why a starter is needed for starting a DC shunt motor?
3. At the time of start, why the field rheostat is kept at minimum resistance point?

3. THEORY

DC machine is a highly versatile energy conversion device and flexible machine. A DC machine is constructed in many forms and for a variety of purposes, from 3 mm stepper motor drawing a few μA at 1.5 V in a quartz crystal watch to a giant 75 MW or more rolling mill motor. It can satisfy the demands of load requiring high starting, accelerating and retarding torques. A DC machine is also easily adaptable for drives with wide range of speed control and fast reversals. These inherent characteristics can be further modified by feedback system. Therefore these are widely used in industry, particularly for tough jobs in spite of their higher initial cost.

At present, the annual production of DC machines is about 40% of the rupee volume in electrical machine production and sales. This is on account of the fact that most vehicles use batteries for storage of electrical energy. In these vehicles and automobiles, DC motors are used as starter, wind shield-wiper, fans and for driving other accessories. For these purposes, almost *millions* of DC motors are built each year.

The design of electrical machines has become a very interesting and challenging topic and is continuously changing with new and improved magnetic, electric and insulating materials, the use of improved heat-transfer techniques, development of new manufacturing processes and the use of computers.

3.1 Construction

The stator of a DC machine consists of:

(*i*) *Yoke or Frame:* This is made up of un-laminated or laminated ferromagnetic material in order to reduce the eddy current loss. The yoke rests on a supporting base.

(*ii*) *Salient Field Poles:* These are bolted to the inner periphery of the yoke. These are made of a stack of steel plates (1 to 1.5 mm thick), riveted together. The pole shoes occupy a major part of the pole pitch. The pole arc is almost 70% of the pole pitch. As a result, the flux density in air around the armature periphery is a *flat-topped* wave. The concentrated field winding or exciting winding, when excited with direct current produces alternate north and south poles, called *hetropolar* construction.

(*iii*) *Bearings, Brushes and Brush Holders:* The stationary carbon brushes are housed in brush holders and are kept pressed on the commutator surface by means of tension controlled spring. Thus the connection of the armature winding to the external circuit is through the commutator and brushes. Brushes are made up of carbon or carbon graphite.

(*iv*) *Interpoles:* These are fixed to the yoke in between the main poles of a DC machine. The interpole winding consists of a few turns of thick wire and is connected in series with the armature so that its magnetomotive force is proportional to the armature current.

(*v*) *Compensating Windings:* These windings are placed in the slot cut in the pole faces similar to the damper winding in case of synchronous machine. Compensating winding is also connected in series with the armature circuit. This winding is, however, used in the large DC machines only.

The rotor of a DC machine consists of:

(*i*) *Armature Core:* It serves two main purposes: (a) housing the armature coils in the slots and (b) providing the low reluctance path to the magnetic flux Φ. It is made from 0.35 to 0.50 mm thick lamination of cold rolled grain oriented silicon steel to keep down the iron losses.

(*ii*) *Armature Winding:* The armature winding is made from copper. It consists of a large number of insulated coils, each coil having one or more turns. These are connected in series or parallel depending upon the type of winding required. There are basically two types of winding: (1) lap and (2) wave. An alternating emf is induced in the coil due to rotation of the armature.

(*iii*) *Commutator:* The two ends of the armature coils are connected to two conducting (copper) segments which are insulated from each other by mica (insulator). Current is collected from the commutator segments by means of carbon brushes (placed over commutator). The function of commutator is to convert armature AC voltage to DC across the brushes, in case of generator (rectifier) and to convert supplied DC voltage across the brushes to AC voltage in the armature, in case of motor (inverter).

(iv) *Shaft:* Armature core and commutator are fixed on the shaft by means of key type arrangement. On both sides of shaft there are bearings which are connected to the yoke.

It may be seen that a DC machine is equivalent to an AC machine plus mechanical converter, i.e., commutator.

3.2 Schematic Diagram of DC Machine

The field winding magnetic axis is called direct axis or D-axis and the axis 90° electrical away from D-axis, is called quadrature axis or Q-axis or interpolar axis or magnetic neutral axis. The magnetic neutral axis (MNA) may be defined as "the axis which is perpendicular to the flux passing through the armature". The armature conductors move parallel to the lines field flux. The emf induced in the conductors or coils along this axis is minimum. Therefore, the brushes are fixed along with this axis for less spark due to commutation. This axis is also called *axis of commutation*. The magnetic neutral axis coincides with the geometrical neutral axis (GNA). The brushes make the connection with the armature winding, hence it may be stated that the armature mmf is always directed along the brush axis.

In Fig. 1.1, the circular symbol represents the armature and two small black rectangles at the opposite side of the circle represent the two brushes.

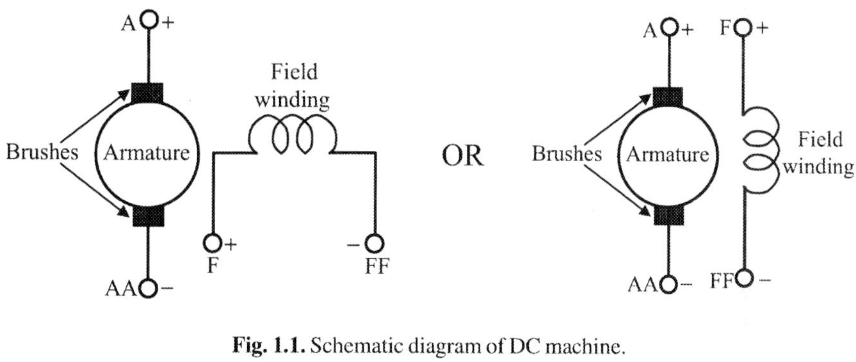

Fig. 1.1. Schematic diagram of DC machine.

3.3 Principle of Operation

The principle of motor action can be stated as follows:

"Whenever a current carrying conductor is placed in a magnetic field, it experiences a force whose direction is given by Fleming's Left Hand Rule".

In 1819, it was discovered by a Danish Physicist, Hans Christian Oersted that an electric current is always accomplished by certain magnetic effects. He found

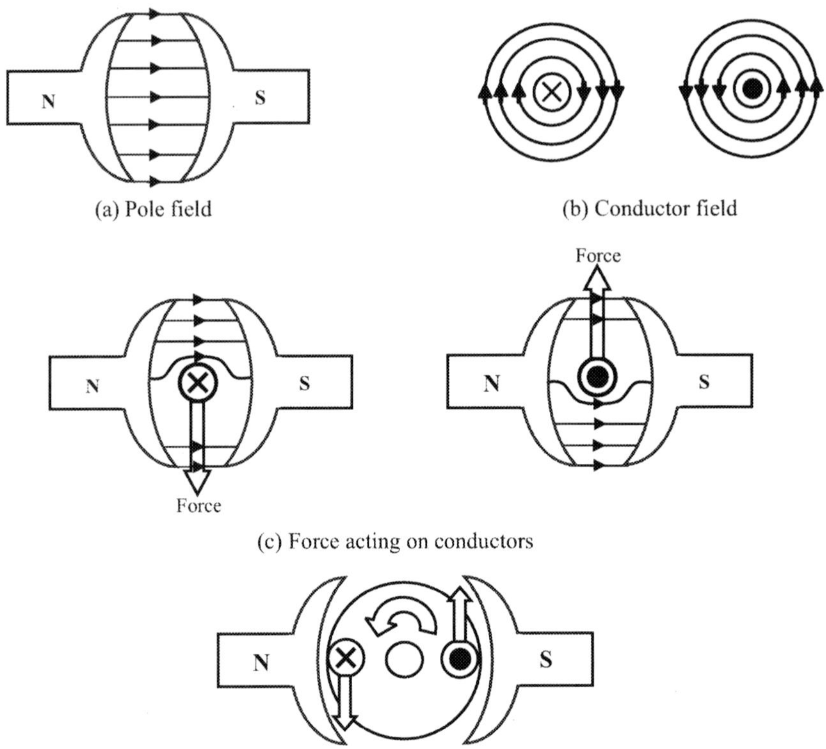

(a) Pole field (b) Conductor field

(c) Force acting on conductors

(d) Rotation of rotor in anti-clockwise direction

Fig. 1.2. The principle of motor action.

that when a current is passed through a conductor placed above the magnetic needle, the needle turns in a certain direction.

This is also called *Motor Rule.* Fig. 1.2 illustrates the principle of motor operation.

Figure 1.2 (a) shows the field set up by the poles. Figure 1.2 (b) shows the conductor field due to the flow of current in the conductors. Figure 1.2 (c) shows the resultant field produced when the current carrying conductor wire is placed in the air gap between the two poles. On the upper side of the cross conductor field lines are additive and on the lower side field lines are subtractive. Thus as per the tension action, a force is applied downward on the conductor. On the upper side of the dot conductor field lines are subtractive and on the lower side field lines are additive. Thus as per the tension action, a force is applied upward on the conductor.

If the cross conductor of the armature is placed under *N* pole and the dot conductor of the armature is placed under *S* pole, then armature rotates in anti-clockwise direction. The same direction can also be obtained by Fleming's Left Hand Rule.

3.4 Equivalent Circuit Diagrams

As the armature rotates, the conductors on it cut the lines of magnetic field in which they revolve, so that an emf is induced in the armature. The polarity of this emf can be obtained by Lenz's law, which states that the direction of an induced emf is such as to oppose the change causing it, which is, of course, the applied voltage V. Therefore, this emf is called *Back EMF* or *Counter EMF* E_b, which can be expressed as

$$E_b = \frac{\Phi ZN}{60}\frac{P}{a}$$

where Φ = field flux, in weber per pole
Z = number of armature conductors
N = Speed in revolution per minute
P = number of poles
a = number of parallel paths in armature circuit.

The value of back emf E_b is always less than the applied voltage V, although the difference is small when the machine is running under no load condition. It the difference between these two quantities which actually drives the current I_a through the resistance r_a of the armature circuit. Thus, we can write an *equation for motor* as $V = E_b + I_a r_a$ and the armature current is $I_a = (V - E_b)/r_a$.

The word back is used because E_b is the voltage at the back of r_a when viewed from armature terminals. Therefore, the generated emf E_b for a motor may also be called *voltage behind armature circuit resistance*.

In DC machines, all parallel paths "a" in the armature are symmetrical. Each path has generated emf E_b and resistance r_p as shown in Fig. 1.3 (a) below. Figure 1.3 (b), (c), (d) and (e) show the equivalent circuits of the armature of a DC machine. The field winding of a DC machine can also be represented in the Fig. 1.3.

3.5 Torque Developed

When the field of a machine is excited and a voltage V is applied to the armature, then the armature current reacts with the field flux to produce the electromagnetic torque T_e, which tends to rotate the rotor. This is essential for the conversion of energy from electrical to mechanical in case of motor. The force on the conductor, $F = B\,I\,L$

where B = flux density, Wb/m^2, I = conductor current, ampere, and L = active length of each conductor, meter,

$\therefore$ Torque developed by each conductor, $T = B\,I\,L \times r$

where r = average radius at which conductors are placed, meter,

Total torque developed by armature, $T_a = BIL \times r \times Z$ Nm

where Z = total number of armature conductors.

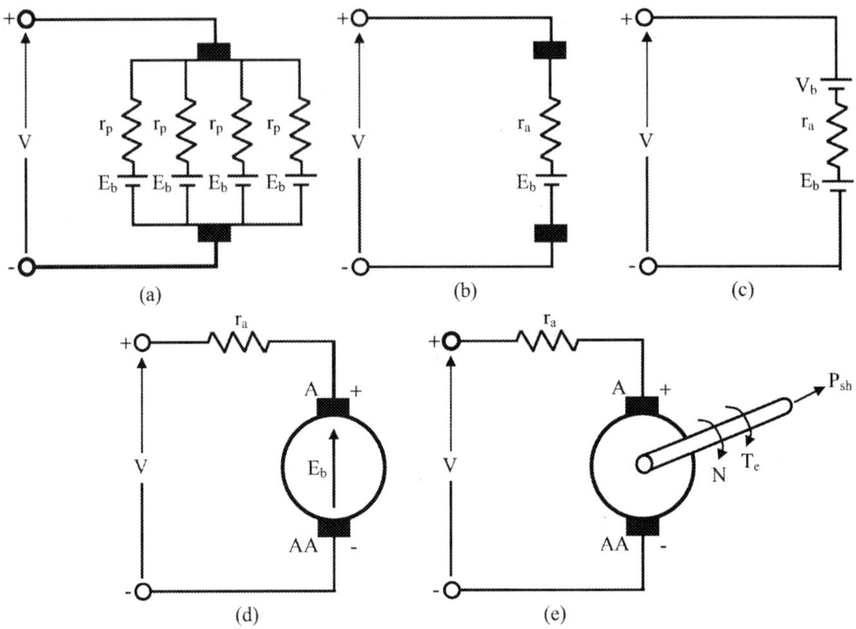

Fig. 1.3. Equivalent circuits pertaining to armature of DC machine.

Now, $B = \Phi/A$, $A = 2\pi\, r\, L/P$, $I = I_a/a$

$$\therefore T_a = \frac{\Phi Z I_a P}{2\pi a} = K_T \Phi I_a \text{ Nm}$$

Thus, the torque is directly proportional to the armature current and field flux or field current.

3.6 Power

The voltage V applied across the armature of motor has to overcome back emf E_b and the armature ohmic drop $I_a r_a$. The voltage equation of motor is

$$V = E_b + I_a r_a$$

Multiplying both sides by I_a, we get

$$V I_a = E_b I_a + I_a^2 r_a$$

This equation is called the *power equation* of DC motor.

Here $V I_a$ is the electrical input to the armature,

$E_b I_a$ is the electrical equivalent of mechanical power P_m developed in the armature causing rotation of the armature.

$I_a^2 r_a$ is the copper loss in the armature.

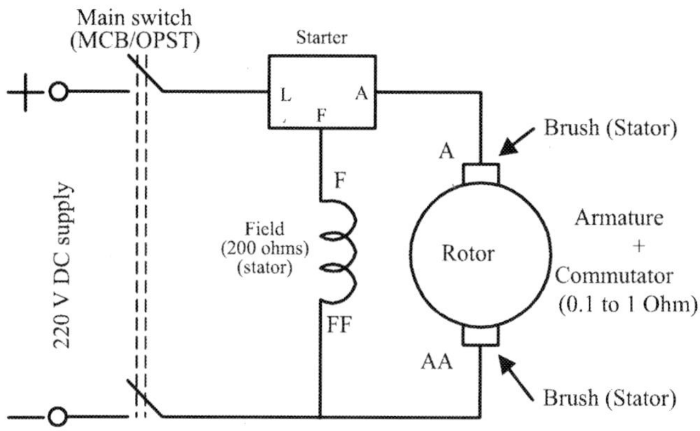

Fig. 1.4. Basic circuit diagram of DC shunt motor.

The useful power is the power available at the shaft, P_{sh}, is less than P_m. This is because P_m has to overcome some mechanical losses (constant losses). Therefore,

$$P_{sh} = P_m - \text{Constant losses} = V I_a - I_a^2 r_a \text{ and}$$

$$T_{sh} = (P_m - \text{Constant losses}) / \omega = (V I_a - I_a^2 r_a) / \omega.$$

The basic circuit diagram of DC shunt motor, with starter, is shown in Fig. 1.4

From the equation of back emf E_b, the equation of speed of the DC shunt motor can be written as

$$N = K_N \frac{V - I_a r_a}{\Phi} \text{ rpm}$$

where K_N = constant for a given motor = $(60 \times a)/(ZP)$,

The speed of the DC shunt motor can be varied by varying

1. The applied voltage V
2. The voltage applied to the armature $V_a = V - I_a r_a$
3. The field flux Φ or field current.

At the time of start, the speed $N = 0$, so $E_b = 0$ and the armature current $I_a = (V - E_b)/r_a = V/r_a$ which is very high and more than rated armature current. Therefore, additional resistance is required to limit the current. This additional resistance is inserted in between line and armature by the use of starter. Hence, STARTER is needed to start the DC shunt motor. During start, move the handle of the starter SLOWLY from off to on. At on position the handle is held up by No Volt release coil. For stopping the motor, simply DISCONNECT the supply. Do not TOUCH the handle. After disconnecting the supply, handle will come automatically to its off position.

When the motor is at rest, high starting torque is required to start the motor. The torque of DC motor is given by $T_e = K_T I_a I_f$. For high starting torque, I_a and I_f should be maximum. I_a is maximum due to zero speed and $E_b = 0$ at start. So keep armature rheostat at MAXIMUM resistance position. In order to make I_f maximum, keep field rheostat at MINIMUM resistance position.

4. APPARATUS REQUIRED

1. DC shunt motor, rating: For example, 5 kW, 220V
2. DC Ammeter: one, 0-2 amps.
3. DC Voltmeter: one, 0-300 volts.
4. Rheostats: two: 370 ohms, 2 amps and 35 ohms, 8 amps
5. Tachometer or speedometer: one.

5. CIRCUIT DIAGRAM

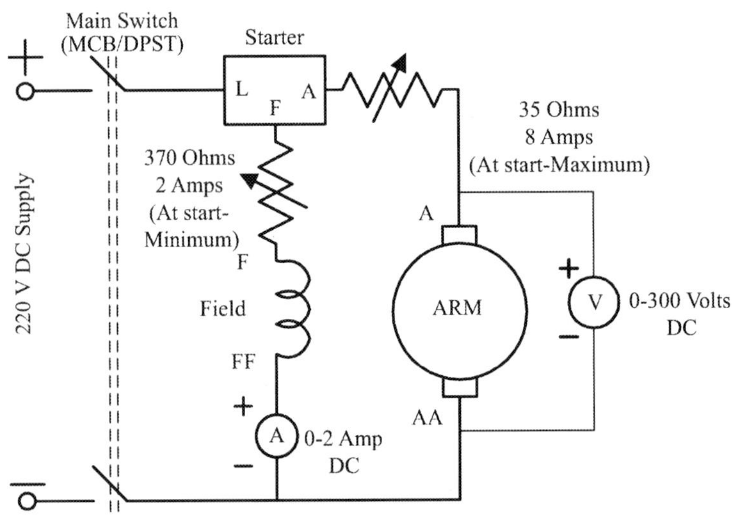

Fig. 1.5. Experimental set-up for speed control of DC shunt motor with armature and field control.

6. PROCEDURE

1. Make the connections as shown in Fig. 1.5.
2. Set up the armature rheostat to maximum value and field rheostat to minimum value.
3. Switch on the DC supply with the help of main switch or MCB. Move the handle of starter ***slowly*** and start the DC motor.

4. Adjust the field current to rated value (say 1 Amp.)
5. Note the voltage across armature and the speed with the help of tachometer.
6. Keeping the field current constant (1 Amp), vary the resistance of armature rheostat to change the armature voltage and note down the speed.
7. Write down the armature voltage and speed in observation Table 1 below.
8. Repeat steps 6 and 7 for another value of field current (0.8 amp)
9. Now keep the armature voltage constant (say 170 Volts) and note down the field current and speed.
10. Keeping the armature voltage constant (170 Volts), vary the resistance of field rheostat to change the field current and note down the speed.
11. Write down the field current and speed in observation Table 2 below.
12. Repeat steps 10 and 11 for another value of armature voltage (say 190 Volts).

7. OBSERVATIONS

Table 1: Armature voltage variation

(1) Field current (constant) I_{f1} = 1 Amp

Sr. No.	Armature Voltage V_a Volts	Speed r. p. m.
1.	150	.
2.	160	..
.	.	...
..	..	....

(2) Field current (constant) I_{f2} = 0.8 Amp

Sr. No.	Armature Voltage V_a Volts	Speed r. p. m.
1.	150	.
2.	160	..
.	.	...
..	..	....

Table 2: Field current variation

(1) Armature voltage (constant) V_{a1} = 170 Volts

Sr. No.	Field Current I_f Amp	Speed r. p. m.
1.	1	.
2.	0.9	..
.	.	...
..	..	....

(2) Armature voltage (constant) V_{a2} = 190 Volts

Sr. No.	Field Current I_f Amp	Speed r. p. m.
1.	1	.
2.	0.9	..
.	.	...
..	..	....

8. RESULTS

The variation of speed with armature voltage and field current has been plotted and studied. The curves are shown in Figs. 1.6 and 1.7 respectively.

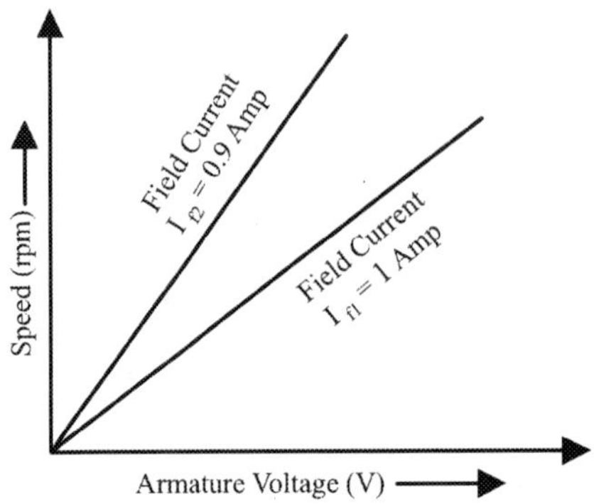

Fig. 1.6. Variation of speed with armature voltage.

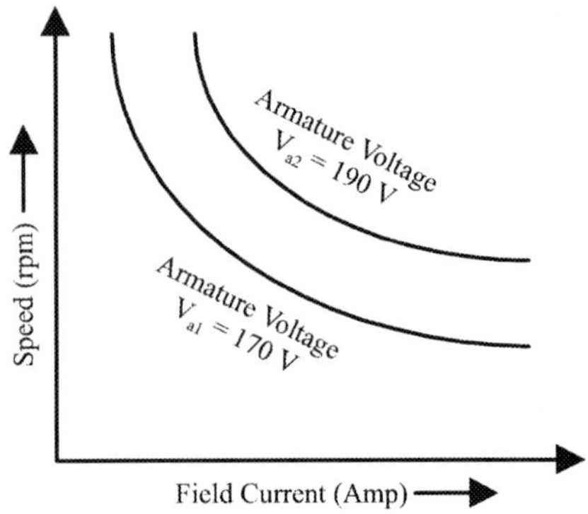

Fig. 1.7. Variation of speed with field current.

9. CONCLUSION

1. The speed of DC shunt motor is directly proportional to the armature voltage ($N \alpha V_a$).
2. The speed is inversely proportional to the field current ($N \alpha 1/I_f$).
3. Thus in order to increase the speed the armature voltage should be increased and / or field current should be decreased.

10. POST-EXPERIMENTAL QUIZ

1. In speed – armature voltage characteristics, why $I_{f2} = 0.9$ A line is above the line of $I_{f1} = 1$ A?
2. In speed –field current characteristics, why $V_{a2} = 190$ V curve is above the curve of $V_{a1} = 170$ V?
3. What is the disadvantage of a 3-point starter?

CHAPTER 2

To Perform Load Test on a DC Shunt Generator

1. OBJECTIVE

The main objective of this experiment is to study the building up of the self-excited DC shunt generator and to see the effects of electrical load on the efficiency and the output quantity, i.e., voltage of the generator.

2. PRE-EXPERIMENTAL QUIZ

1. What is the essential condition for building up of self-excited DC shunt generator?
2. How mechanical power is converted into electrical power in case of a generator?
3. What do you mean by "**critical resistance**"?

3. THEORY

An electromechanical energy conversion device converts electrical energy into mechanical energy, i.e., motor and mechanical energy into electrical energy, i.e, generator. Electromechanical energy conversion takes place via the medium of a magnetic or electric field. The magnetic field being the most suited for practical conversion device. In case of motor, the output is available at the shaft, i.e., speed. Some of the applications are pumps, fans, mixer, lathe machine, drilling, cutting, etc. In case of generator, the mechanical energy is given by any motor (AC or DC) or any turbine (steam or water). Even though a generator converts mechanical energy into electrical energy, little amount of electrical energy is required for the operation. This energy is in the form of excitation, flux or field current. This excitation can be separate excitation (independent source) or self excitation (residual magnetism). This little amount of electrical energy is amplified with the help of speed and number of armature conductors.

In case of DC generation, the same DC motor can be operated as a DC generator. In rotating electrical machines, the output voltage or emf can be generated in the armature windings by:

1. rotating these windings through a magnetic field;
2. rotating the magnetic field with respect to these windings; and
3. designing the magnetic circuit to have variable reluctance with rotor rotation.

Consider the generation of emf due to the relative motion between field flux and armature winding. Elementary form of DC generator is shown in Fig. 2.1.

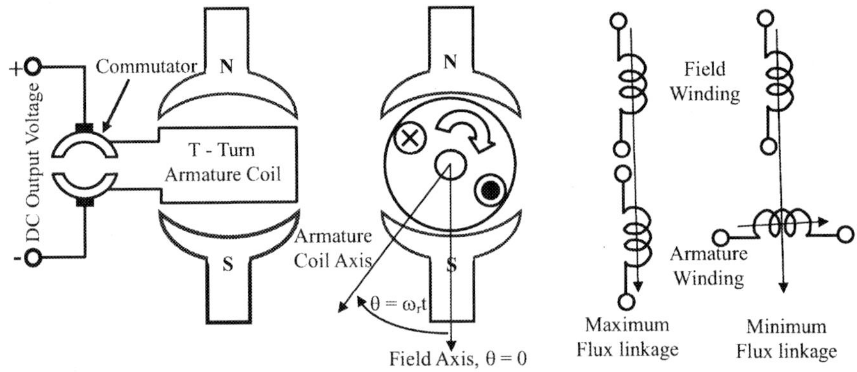

Fig. 2.1. Generation of emf.

The flux density wave in the air gap is assumed to be sinusoidal. ω_r is the *angular velocity in radian per second.* At the time of origin $t = 0$, time angle $\theta = \omega_r t = 0$ and the field winding axis and armature coil axis are coincident. The flux passing through or linking the armature coil is maximum ($= \Phi \cos 0$). When the time angle $\theta = \omega_r t = 90°$, the flux passing through or linking the armature coil is zero ($= \Phi \cos 90$). For $\theta = \omega_r t = 180°$, the flux passing through or linking the armature coil is again maximum but in reverse direction ($= \Phi \cos 180$). In view of this, an expression giving variation of the flux passing through the armature coil can be written as $\Phi \cos \theta$ or $\Phi \cos \omega_r t$.

Therefore, the flux linkages ψ with the armature coil at any time t are

$\psi = T$ (Flux linkage passing through the armature coil)

$$= T\,\Phi \cos \omega_r t \tag{1}$$

By Faraday's law, the emf induced in the T turn coil is

$$e = \frac{d\psi}{dt} = \underbrace{T\,\Phi\omega_r \sin\omega_r t}_{I} \underbrace{-T\frac{d\Phi}{dt}\cos\omega_r t}_{II} \tag{2}$$

The first term is the speed voltage because it contains ω_r speed in it. The second term is the transformer voltage because it involves the time derivative of the flux

Φ, i.e., $d\Phi/dt$. Here the flux is time invariant (i.e., Φ is not varying with time due to DC supply), and $d\Phi/dt = 0$.

Therefore, the generated emf is

$$e = T\Phi\omega_r \sin\omega_r t \tag{3}$$

The voltage appearing at the brushes is unidirectional due to the commutator and its average or DC value is

$$E_{av} = \frac{1}{\pi}\int_0^{\pi} T\Phi\omega_r \sin\omega_r t\, d(\omega_r t)$$

$$= \frac{2}{\pi} T\Phi\omega_r \tag{4}$$

Putting the expressions for $\omega_r = 2\pi f = 2\pi \cdot Pn/2 = \pi Pn$ and $T = Z_T/2$, Z_T is the total number of conductors in a coil, then

$$E_{av} = Z_T\, n P\Phi \tag{5}$$

In armature winding, there are parallel paths "*a*" with equal number of conductors, so write down average voltage per conductor.

$$E_{av} \text{ per conductor} = \Phi n P$$

Now, if Z is the total number of armature conductors arranged in "*a*" parallel paths as shown in Fig. 2.2, then series conductors between brushes are Z/a. Thus, the voltage appearing between the brushes or between the terminals *A* and *AA* is

E_a = (E_{av} per conductor) × (number of conductors in each parallel path)

$$E_a = \frac{\Phi Z n P}{a} \tag{6}$$

For a particular machine, $Z\,P/a$ = constant.

$$\therefore E_a = K_a \Phi n \tag{7}$$

Thus E_a is the induced emf in the armature winding with parallel paths "*a*". In motor this voltage is called *back emf* E_b. The parallel paths are equal to the number

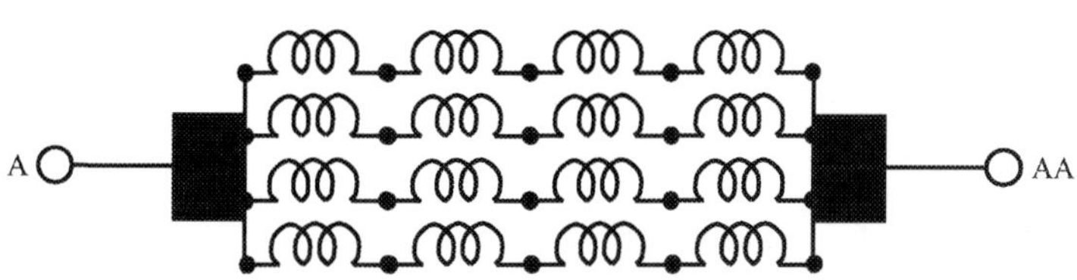

Fig. 2.2. Parallel connections of coils in armature winding.

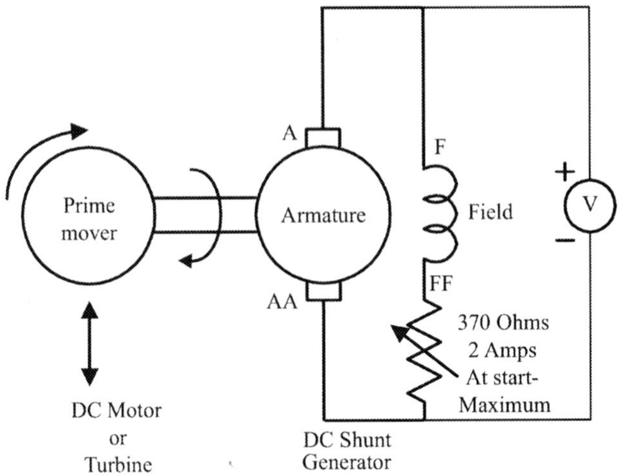

Fig. 2.3. Basic diagram of DC shunt generator.

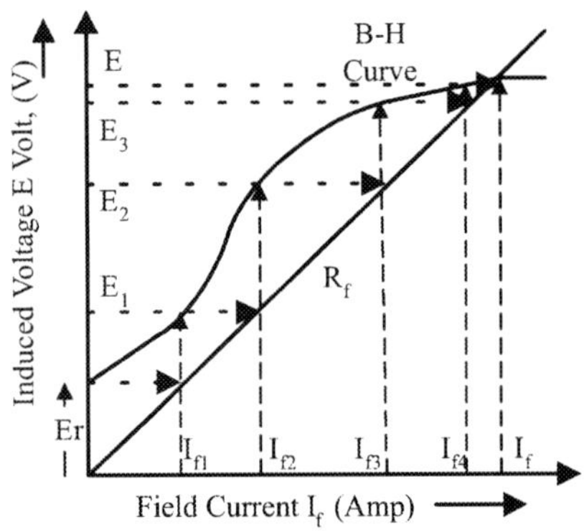

Fig. 2.4. Build-up of DC shunt generator.

of poles in case of lap winding and equal to 2 in case of wave winding. It is observed that E_a depends upon the flux per pole Φ and not upon the shape of the flux density wave. For the same value of peak flux density, the average value of brush voltage would be more for a flat topped flux density wave. It is because of this reason that DC machines are designed with flat topped flux density wave.

For a generator, running at constant speed, the above expression can be written as

$$E_a = K_1 \Phi \text{ or } E_a = K_2 I_f \tag{8}$$

Thus the voltage generation in the armature winding is directly proportional to field flux or field current. The basic diagram of DC shunt generator is shown in Fig. 2.3 and voltage building up of the same generator is shown in Fig. 2.4.

DC shunt generator is a self excited generator. Its armature provides the current to the field circuit and the field current establishes the magnetic flux in which armature rotates. In DC shunt generator an expression for the voltage generated in the armature winding is given by equation 6.

With the generator driven at constant rated speed, there is small generated emf, E_r due to residual magnetism. The residual magnetism is the most ***ESSENTIAL*** condition for obtaining the generated voltage (i.e., for the build up of voltage). The residual magnetism is the remaining of small flux in the field pole or in the field winding structure due to the machine excitation during previous usage of the machine. This residual magnetism is due to hysteresis effect. This flux is present when the field current I_f is zero.

Due to this voltage E_r, a current of magnitude $I_{f1} = (E_r / R_f)$ flows in field winding. Due to this I_{f1}, the generated emf increases to E_1, if the flux due to I_{f1} and residual flux are in the same direction. Due to this E_1, the field current increases to $I_{f2} = (E_1 / R_f)$ as shown in Fig. 2.4. This current increases the flux and hence armature voltage goes up to value E_2. This process continues till the point of intersection of two curves (Field Resistance line and B-H curve) is reached. At this point, the voltage is E_a which circulates I_f through the field winding. The induced voltage cannot increase further.

If total resistance of field winding and field rheostat is more than the critical field resistance R_{fc}, then there is no building up of the generator. For building up of the generator, the total resistance should be less than the critical field resistance, i.e., $R_f < R_{fc}$.

At the time of start, in order to maintain the generated emf less than the rated voltage, the field rheostat should be kept at maximum position. If the resistance of field rheostat is minimum, then the field current is maximum and the generated voltage may be more than the rated voltage, which may damage the insulation of the winding.

After building up of voltage E_a, the load test is performed to know **how its terminal voltage changes with the electrical load.** It is generally assumed that the speed of the generator is maintained constant by prime mover. During the test if the prime mover speed falls due to loading, it has an additional effect on the generator's performance. The other purpose of the test also could be **to find its regulation, efficiency and to measure its temperature rise at different loads.**

4. APPARATUS REQUIRED

1. Test machine - Generator, Rating: 5 kW, Speed: 1000 rpm, 220 Volts, etc.
2. Driving machine – DC Shunt Motor, Rating: 5 kW, Speed: 1000 rpm, 220 volts, etc.
3. Field rheostat: Two, 300 ohms, 2 amps.
4. DC Ammeter: Two, 25-50 amps.
5. DC Voltmeter: Two, 0-300 volts.
6. Loading rheostat: 5 kW, 230 volts.
7. Tachometer.

5. CIRCUIT DIAGRAM

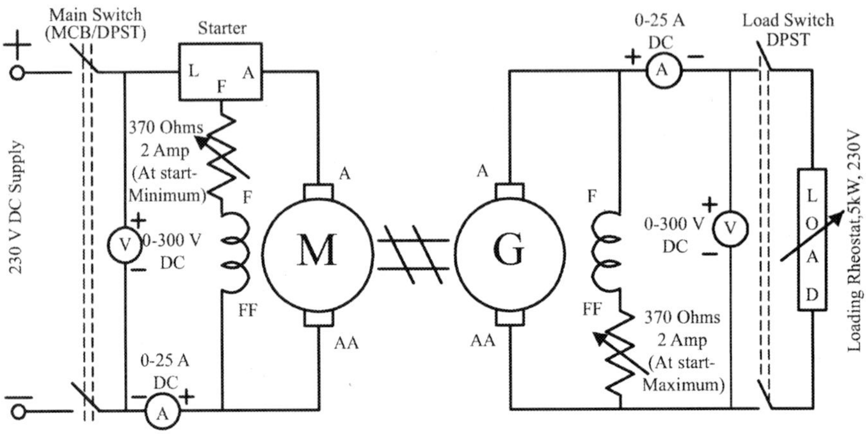

Fig. 2.5. Experimental set-up for load test on DC shunt generator.

6. PROCEDURE

1. Make the circuit connections as shown in Fig. 2.5.
2. Set up the field rheostat of motor to minimum value and field rheostat of generator to maximum value. Keep the loading rheostat in off position.
3. Switch on the DC supply with the help of main switch or MCB. Move the handle of starter ***slowly*** and start the DC motor.
4. Adjust the speed of DC motor with the help of motor field rheostat to the rated speed of generator. (Keep the speed ***constant*** throughout the experiment with the help of motor field rheostat).
5. Decrease the resistance of field rheostat of generator to build up the rated terminal voltage of generator. The field rheostat adjusted at this position should not be disturbed for this set of reading.
6. Note down the corresponding readings on motor side, i.e., motor voltage, current and the corresponding readings on generator side, i.e., generator voltage, current.
7. Increase the load current by switching on one switch (1 amp) of loading rheostat and repeat the step 6.
8. Increase the load current by switching on another switch of loading rheostat and repeat the step 6.
9. Repeat the step 8 up to rated current of generator.

7. Observations

Speed = ... r. p. m. (Constant)

Observations					Calculations		
Sr. No.	Motor side		Generator side		Motor I/P $M_i = V_m \times I_m$ Watts	Gen O/P $G_o = V_g \times I_g$ Watts	Gen Efficiency $= \sqrt{\left(\frac{G_0}{M_i}\right)}$
	Voltage V_m	Current I_m	Voltage V_g	Current I_g			
1.	.	.	.	1A			
2.	..	..	..	2A			
3.	...	...	...	3A			
.	....	....	....	.			
..	.....	.....	.....	..			

Note: It is assumed that both the machines coupled have approximately the same efficiency.

So $$\text{Generator efficiency} = \frac{\text{Output of generator}}{\text{Input of generator}}$$

$$\text{Motor efficiency} = \frac{\text{Output of motor}}{\text{Input of motor}}$$

As the output of motor = Input of generator

$$\eta_G \times \eta_M = \frac{\text{Output of generator} \times \text{Output of motor}}{\text{Input of generator} \times \text{Input of motor}} = \frac{\text{Output of generator}}{\text{Input of motor}}$$

$$\eta_G = \sqrt{\frac{\text{Output of generator}}{\text{Input of motor}}}$$

8. RESULTS

The external characteristic of DC shunt generator and efficiency curve are plotted and studied. The characteristics are shown in Figs. 2.6 and 2.7 respectively.

9. CONCLUSION

1. The terminal voltage reduces as the load or load current is increased.
2. There are three reasons of voltage drop, due to load, in the DC shunt generator: (i) due to armature reaction, (ii) armature resistance drop, and

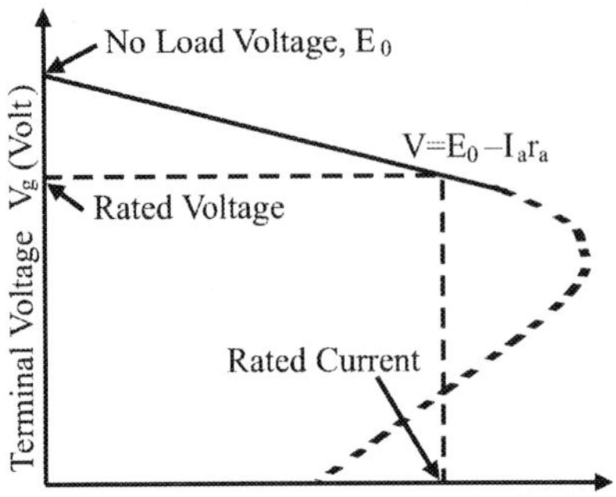

Fig. 2.6. External characteristic of a DC shunt generator.

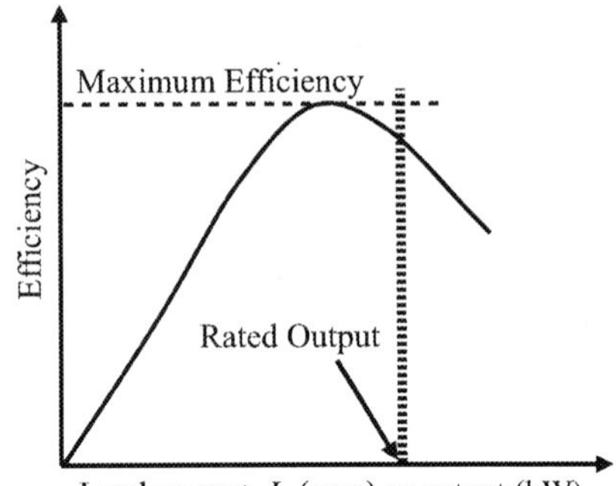

Fig. 2.7. Output vs efficiency curve of a DC shunt generator.

(iii) reduction in field current due to a reduction in armature voltage due to (i) and (ii) above. This is known as external characteristic.

3. The internal characteristic is the characteristic between voltage and load current when armature reaction effect is considered.
4. The efficiency is zero at no load. It increases as the load increases and is maximum when variable losses are equal to constant losses.
5. The efficiency at rated output is less than the maximum value and the operating point is after maximum efficiency.

10. POST-EXPERIMENTAL QUIZ

1. What do you mean by residual magnetism?
2. What are the reasons for voltage drop in the DC shunt generator?
3. The efficiency at rated output is less than the maximum value and the operating point is after maximum efficiency. Why?
4. Explain armature reaction.

CHAPTER 3

To Perform Open Circuit and Short Circuit Test on a Single-Phase Transformer

1. OBJECTIVES

The main objectives of this experiment are:

1. to obtain core loss and copper loss at rated voltage and current conditions.
2. to determine the voltage transformation ratio constant.
3. to determine the values of parameters of equivalent circuit diagram of a transformer.

2. PRE-EXPERIMENTAL QUIZ

1. Explain transformer.
2. How are the emfs induced in primary and secondary windings?
3. What are the functions of no load current?
4. What is the ratio of no load current to full load current? How does it compare with the same in case of an induction motor?

3. THEORY

A transformer is a static electromagnetic device or an alternating current device used for changing a given alternating voltage to another of different magnitude and of same frequency, f Hz. It transfers electrical energy from one circuit to another through the medium of magnetic field and without making electrical connection so it is used to isolate the electrical circuit. It also transforms current and impedance.

The transformer is one of the most important components of a electrical power system ranging from low power to very high power (MW). Transformers are built in a small range of sizes from tiny units used in electronics and communication systems to big sizes used in high voltage electrical systems, weighing hundreds of tons. A circuit model and performance analysis of transformer is necessary for understanding of many electronic systems, control systems and all power systems.

Transformers are used as AC power systems because they make possible power generation at the most desirable and economical level (10-20 kV), power transmission at an economical transmission voltage (as high as 400 – 1000kV) and power utilization at most convenient distribution voltage (230 / 440V). The voltage level of 3.3, 6.6 and 11 kV for large motors in industrial applications can also be obtained by transformer.

In a transformer, the electrical energy transfer from one circuit to another circuit takes place without the use of moving parts. Therefore, it has the highest possible efficiency as compared to all other electrical machines and requires less maintenance and supervision.

3.1. Construction

The simplest form of transformer consists of two coils or windings which are mutually coupled, through a ferromagnetic core, as shown in Fig. 3.1. The magnetic core is a stack of thin silicon steel laminations about 0.35 to 0.5 mm thick. In order to reduce the eddy current losses, these laminations are insulated from each other by thin layers of varnish or paint. For reducing the core losses, nearly all transformers are made from cold-rolled grain oriented (CRGO) silicon steel sheet. This material, when manufactured under cold condition produces the grain in proper direction in the rolling process in order to have low core losses and high permeability. The steel sheets or laminations used are of I, L, U, T or E shape. The vertical portions of the core are called *limbs* or *legs* and the top or bottom portions are called *yoke*.

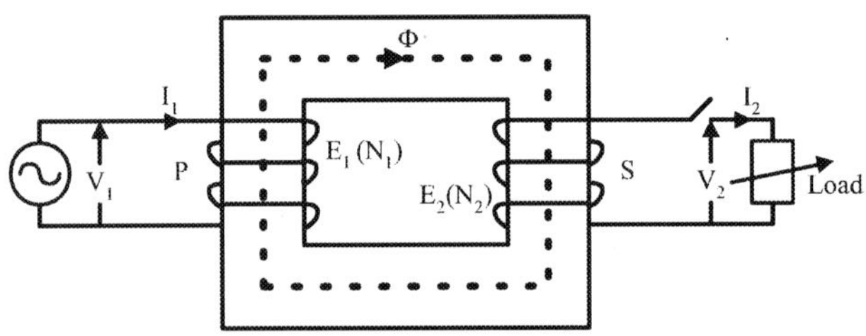

Fig. 3.1. Schematic diagram of a two-winding transformer.

The windings are placed over the limbs. These are made of circular, square or rectangular copper or aluminum conductors. The current capacity of the windings depends upon the cross-section of the conductors. The winding should be mechanically stable in respect to the forces developed when there is sudden short circuit. Low voltage (LV) winding is placed near to core to reduce the thickness of insulation. The LV and high voltage (HV) windings can be divided into sections and placed one over the other to have better operation. The types of windings used in transformers are: (i) cylindrical, (ii) helical, (iii) cross-over, (iv) disc and (v) sandwich windings.

3.2. Types

Depending upon the construction, there are two general types of transformers: (i) *core type* and (ii) *shell type*. These two types depend on the manner in which

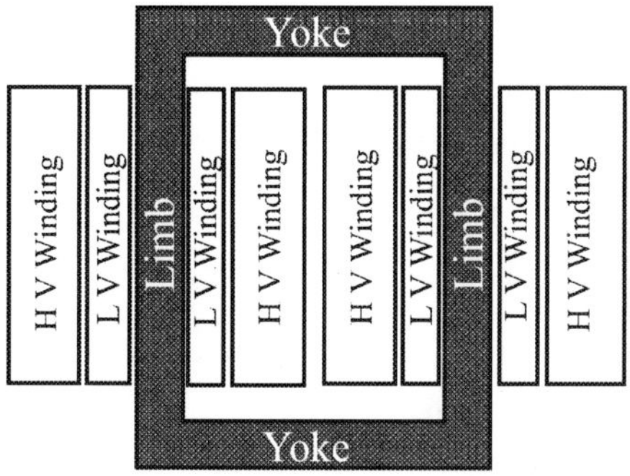

Fig. 3.2. Single phase core type transformer.

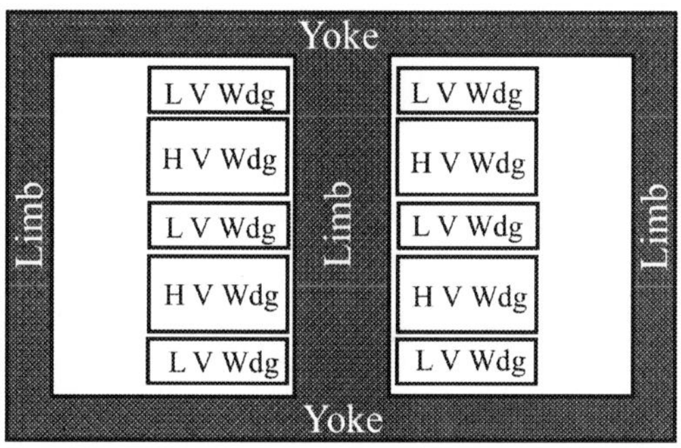

Fig. 3.3. Single phase shell type transformer.

the windings are wound around the core. In the core type, the windings surround a considerable part of core (limbs), as shown in Fig. 3.2. In the shell type, core (outer limbs) surrounds a major part of the windings as shown in Fig. 3.3. The winding is wound on the central limb. Thus for single phase transformers, core type has two legged core whereas the shell type has three legged core.

Depending upon the application in power system, there are two types of transformers: (i) *Power transformer* and (ii) *Distribution transformer*.

Power transformers are of generally high rating transformer and distribution transformers are of low rating transformer. Power transformers are used in generating stations or substations at each end of transmission line for stepping up or stepping down the voltages. They are connected when there is load and disconnected when there is little load. They are designed to have maximum efficiency near to full load and to have greater leakage reactance. For this transformer the voltage regulation is less important than limiting the current due to higher leakage reactance.

Distribution transformers are used to step down the voltage suitable to appliances and loads, used by consumers. In order to maintain the consumer voltage constant, they are designed to have less leakage reactance. They are kept in operation for 24 hours whether there is load or no load. Therefore, they are designed to have maximum efficiency near to 50% of full load. The output voltage of this transformer is 440 V (line) for domestic consumers.

When the transformers are used to measure the voltage (PT) and current (CT), then these are called *instrumentation transformers.*

When the transformers are used to insulate two circuits, then these are called *insulation transformers.*

3.3 Operation

The circuit symbol for a two-winding transformer is shown in Fig. 3.4. A coil represents winding of a transformer. The two vertical lines represent the insulated magnetic material and strong magnetic coupling between the two windings.

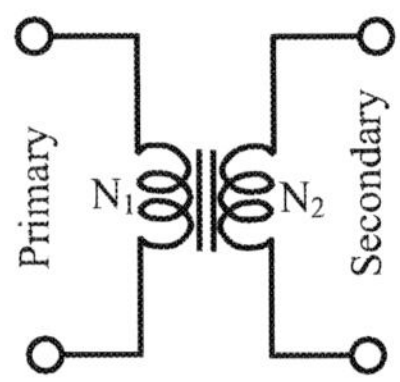

Fig. 3.4. Circuit symbol for a two-winding transformer.

The winding, which is connected to an alternating source voltage, V_1, or the winding which receives energy from the source of AC supply is called *primary winding, P*. Due to this voltage, an alternating no load current I_0 starts flowing through N_1 turns. The alternating mmf $N_1 I_0$ sets up an alternating flux Φ which is confined to the high permeability iron path. The amplitude *of* this flux Φ in the core depends on primary voltage, V_1, frequency, *f* and number of turns, N_1. This flux links the other winding, which is called *secondary winding, S*. The winding which is connected to the load and delivers energy is called secondary winding, *S*. A transformer works on the principle of electromagnetic induction. According to this principle, an emf is induced in a coil if it links a changing flux. A voltage is induced in this secondary winding of the same frequency as the primary voltage but its magnitude depends on the number of secondary turns, N_2. From this winding, the output is taken.

If load is connected to the secondary winding, then an alternating current flows through it and energy is delivered to the load. Thus electrical energy is transferred from primary to secondary by a common magnetic circuit. Since there is no relative motion between the windings, frequency of the induced emf in secondary is exactly same as the frequency of the applied voltage to primary.

Let the sinusoidal flux be given by

$$\Phi = \Phi_{max} \sin \omega t \qquad (1)$$

where Φ_{max} is the maximum value of core flux.

As per Faraday's law, the instantaneous emf e_1 is induced in primary winding and emf e_2 is induced in secondary winding. The expression for emf e_1 is given by

$$e_1 = -N_1 \frac{d\Phi}{dt} = -N_1 \frac{d\Phi_{\max} \sin \omega t}{dt} = -N_1 \Phi_{\max} \omega \cos \omega t = N_1 \Phi_{\max} \omega \sin(\omega t - \pi/2) \tag{2}$$

$$= E_{1\max} \sin(\omega t - \pi/2) \tag{3}$$

$$\therefore rms\, E_1 = \frac{E_{1\max}}{\sqrt{2}} = \frac{N_1 \Phi_{\max} 2\pi f}{\sqrt{2}} = 4.44 f N_1 \Phi_{\max} \tag{4}$$

Similarly, the induced emf E_2 in rms is expressed by

$$rms\, E_2 = \frac{E_{2\max}}{\sqrt{2}} = \frac{N_2 \Phi_{\max} 2\pi f}{\sqrt{2}} = 4.44 f N_2 \Phi_{\max} \tag{5}$$

Equations 4 and 5 are called *emf equations of a transformer*. The winding having higher number of turns, has high voltage and is called *high voltage (HV) winding*. The winding having lower number of turns, has low voltage and is called *low voltage (LV) winding*.

3.4 Voltage Ratio

The expression for volts per turn is E/N.

The primary volts per turn is $E_1/N_1 = 4.44\Phi_{max} f$ (6)

The secondary volts per turn is $E_2/N_2 = 4.44\Phi_{max} f$ (7)

Equations 6 and 7 show that the volts per turn are same in both the windings. Therefore,

$$\frac{E_1}{N_1} = \frac{E_2}{N_2} \text{ and } \frac{E_2}{E_1} = \frac{N_2}{N_1} = k$$

where k = voltage transformation ratio or turns ratio.

There is a small voltage difference between V_1 & E_1 and E_2 & V_2.

$$\therefore \frac{V_2}{V_1} = \frac{N_2}{N_1} = k \tag{8}$$

If k is more than 1, then output (secondary) voltage is more than the input (primary) voltage and the transformer is called ***step-up transformer***.

If k is less than 1, then output (secondary) voltage is less than the input (primary) voltage and the transformer is called ***step-down transformer*** (distribution transformer).

If k is equal to 1, then output (secondary) voltage is equal to the input (primary) voltage and a transformer is called ***one to one (1:1) transformer*** (insulation transformer).

3.5 Ideal Transformer

An ideal transformer is an imaginary (hypothetical) transformer and has the following properties:

1. The primary and secondary windings have no resistance (no copper loss and no resistive voltage drop).
2. There is no leakage flux and no leakage reactances so that the entire flux is confined to the core and links both the windings.
3. The core has infinite permeability so that zero magnetizing current is needed to establish flux in the core.
4. The core loss (hysteresis and eddy current) is zero and the efficiency is 100%.

For an ideal transformer, $V_1 = E_1$ and $V_2 = N_2$. As a load is connected (see Fig. 3.1) to the secondary winding, a current I_2 flows, decided by the load impedance Z_L, such that $I_2 = E_2/Z_L$. This is due to E_2 and E_2 is due to the induction. According to the Lenz's law, the induced voltage is in such a direction that it opposes its cause. In this case, E_2 is induced due to a changing flux Φ. Thus, the direction of E_2 is such that a current I_2 due to it tries to reduce the changing flux Φ. But, for constant V_1, Φ_{max} is constant. These two conditions are satisfied only when the primary winding automatically draws a proper current I_1, from the supply, so that secondary mmf ($= I_2N_2$) is neutralized. This balance of ampere-turns on the two sides is another operating principle of the transformer. From this it can be stated that the demagnetizing ampere-turns of the secondary are equal and opposite to the magnetizing ampere-turns of the primary. Therefore for ideal transformer

$$N_1I_1 = N_2I_2 \therefore \quad \frac{I_1}{I_2} = \frac{N_2}{N_1} = k = \frac{V_2}{V_1} \tag{9}$$

This implies that the current ratio is *in inverse* of voltage ratio. Equation 8 can be written as

$V_1I_1 = V_2I_2$ or input volt-amperes = output volt-amperes

Thus, input kVA is equal to output kVA.

For drawing the phasor diagram of an ideal transformer at no load, consider equation (1) at time, $t = 0$. The flux Φ is zero, so it is shown by drawing a horizontal line, as shown in Fig. 3.5. (a). This flux is set up by voltage V_1, so it should lead the flux by 90°. The exciting current I_e is drawn lagging by 90° to the voltage V_1 and in phase with flux since the primary winding is a inductive coil. The emfs E_1 and E_2 are induced due to this flux. So these emfs are drawn lagging by 90° to the flux.

For drawing the time diagram of an ideal transformer at no load, first flux is drawn sinusoidally varying as shown in Fig. 3.5. (b). The exciting current I_e in phase with the flux, voltage V_1 leading by 90°, E_1 and E_2 lagging by 90° to the flux.

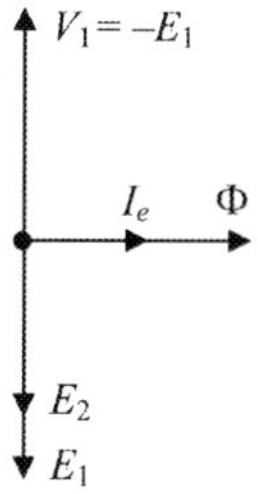

(a) Phasor diagram

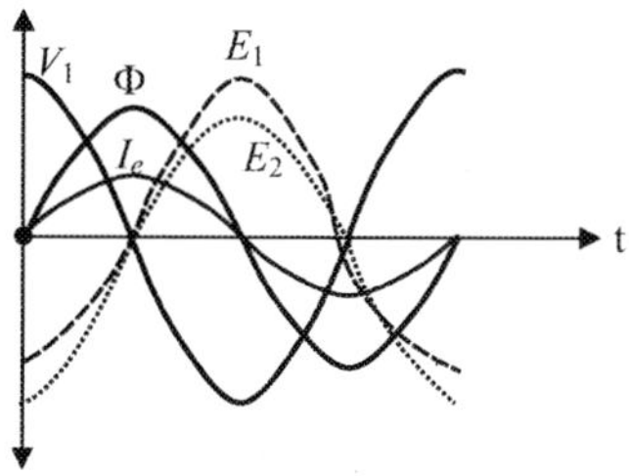

(b) Time diagram

Fig. 3.5. Ideal transformer.

3.6 Transformer on No Load

In this case, for simplicity neglect the resistances of the windings. A transformer is said to be on no load if the secondary winding is open circuited and the primary is connected to a source of constant sinusoidal voltage of frequency *f*. The secondary current is zero and a small amount of primary current I_o flows in the primary, which is called *no-load current*. This current consists of two components: (i) magnetizing current or reactive component or wattless component of current, I_μ or I_m or I_r, and (ii) core loss component or active component or working component or wattful component of current, I_l or I_w or I_i or I_c or I_a.

The magnetizing current produces the flux in the core and is therefore is in phase with the flux, Φ. The core loss component is responsible for hysteresis and eddy current losses in the core and is in phase with the allied voltage, V_1. The no load current is 2 to 6% of the rated current of the transformer.

The phasor diagram under no-load condition is shown in Fig. 3.6. (a) with power factor angle θ_0. cos θ_0 is no-load power factor and its value is near to 0.1, since the no load current phasor is more inclined to flux axis. The equivalent circuit diagram is shown in Fig. 3.6 (b). The actual transformer is now replaced by an ideal transformer with a resistance R_o or R_c and an inductive reactance X_o or

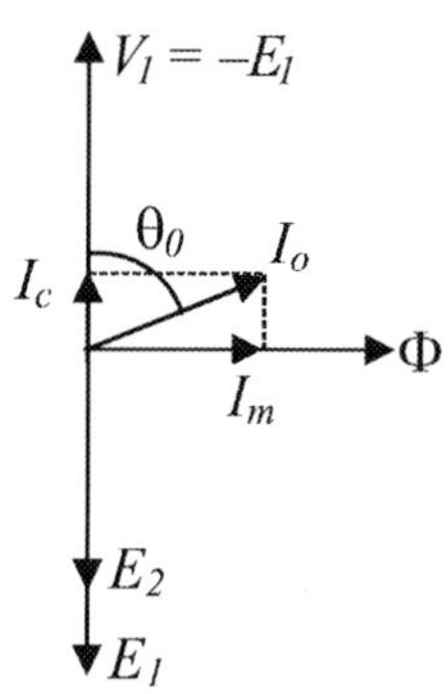

(a) Phasor diagram

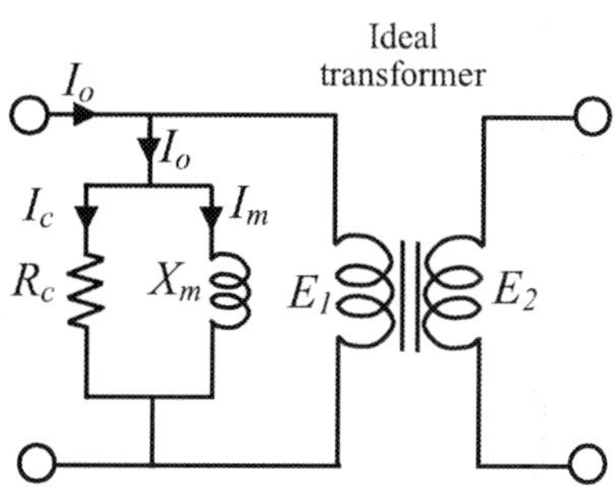

(b) Equivalent circuit diagram

Fig. 3.6. No-load transformer.

X_m in parallel. R_0 represents the core loss which takes I_c. An inductive reactance X_m takes I_m.

$$I_c = I_0 \cos \theta_0,\ I_m = I_0 \sin \theta_0,\ I_0 = \sqrt{I_c^{\ 2} + I_m^2}\text{ , power factor} = I_c / I_0$$

Core loss = active power = V, $I_c w$, reactive power $V_1 I_m V_{Ar}$

3.7 Real or Practical Transformer

(a) With Resistances

The assumptions made for ideal transformer are not valid for a practical transformer. In case of a practical transformer, primary and secondary windings have some resistances r_1 and r_2 respectively. It can be represented in the equivalent circuit diagram by drawing resistances r_1 and r_2 in series with the respective winding. This is shown in Fig. 3.7.

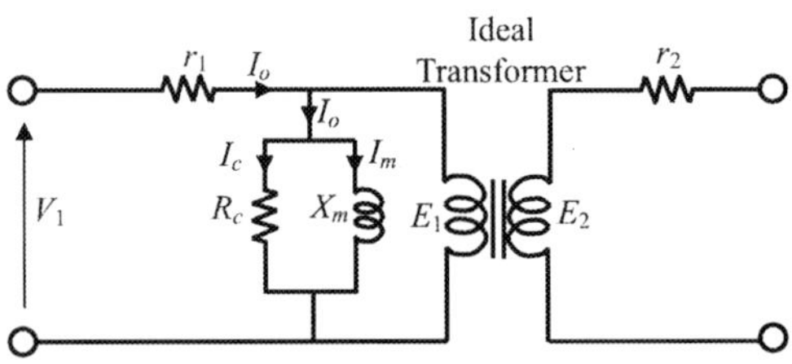

Fig. 3.7. Equivalent circuit diagram of real transformer.

(a) With Leakage Reactances

All the flux produced by primary winding does not remain in the core. A small amount of flux is completing its path through the nearby air. This flux is known as *primary leakage flux* and is denoted by Φ_{L1} as shown in Fig. 3.8. This flux links primary winding only and induces an emf E_{L1}.

The major portion of flux produced secondary current I_2, opposes the main flux Φ, and small portion of secondary current flux completes its path through

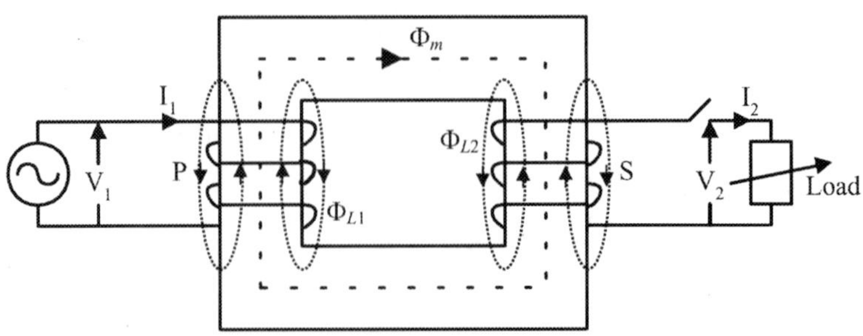

Fig. 3.8. Transformer windings leakage fluxes.

nearby surrounding air. This small portion of flux is called *secondary leakage flux* and is denoted by Φ_{L2} as shown in Fig. 3.8. This flux links secondary winding only and induces an emf E_{L2}. The flux which passes through the core and links both the windings is called mutual flux Φ_m.

Voltages E_1 and E_2 are induced due to mutual flux Φ_m. The voltages E_{L1} and E_{L2} are induced due to *leakage fluxs* Φ_{L1} and Φ_{L2} respectively. Therefore, the effect of leakage flux is equivalent to an inductance in series with each winding such that the voltage drop in each series inductance is equal to that produced by the leakage flux. In other words, the leakage flux can be represented in the equivalent circuit by drawing leakage inductive reactances x_1 and x_2 connected in series with primary and secondary windings respectively.

3.8 Equivalent Circuit Diagrams

The *exact* equivalent circuit diagram with all variables and parameters, is shown in Fig. 3.9. Let us consider that r_2' is the secondary winding resistance reflected or transferred to the primary side. This resistance should produce the same effect in the primary as it produces in the secondary. Therefore the power consumed by r_2' due to I_1' should be equal to the power consumed by r_{2} due to I_2.

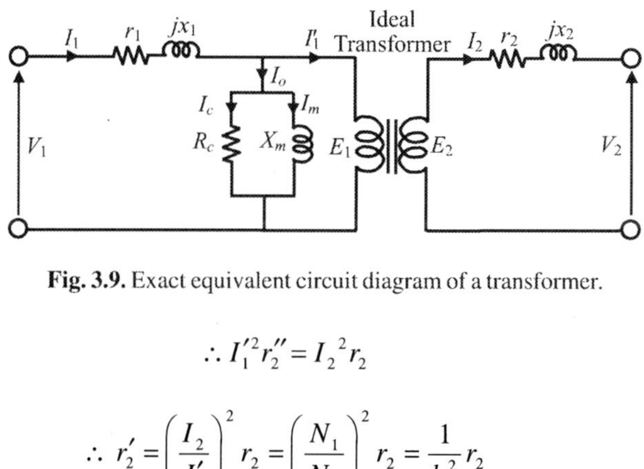

Fig. 3.9. Exact equivalent circuit diagram of a transformer.

$$\therefore\ I_1'^2 r_2'' = I_2^2 r_2 \tag{10}$$

$$\therefore\ r_2' = \left(\frac{I_2}{I_1'}\right)^2 r_2 = \left(\frac{N_1}{N_2}\right)^2 r_2 = \frac{1}{k^2} r_2 \tag{11}$$

Similarly, let us consider that x_2' be the secondary winding reactance reflected or transferred to the primary side. This reactance should produce the same effect in the primary as it produces in the secondary. Therefore the reactive power absorbed by x_2' due to I_1' should be equal to the reactive power absorbed by x_{2} due to I_2. The reactive power is VI sin θ = $IZ.I.X/Z = I^2X$.

$$\therefore\ I_1'^2 x_2' = I_2^2 x_2 \tag{12}$$

$$\therefore\ x_2' = \left(\frac{I_2}{I_1'}\right)^2 x_2 = \left(\frac{N_1}{N_2}\right)^2 x_2 = \frac{1}{k^2} x_2 \tag{13}$$

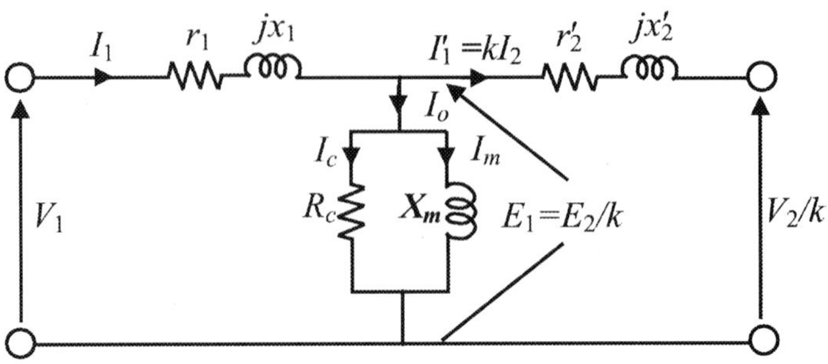

Fig. 3.10. Exact equivalent circuit diagram of a transformer referred to primary.

Thus, the exact equivalent circuit diagram referred to primary can be drawn as shown in Fig. 3.10.

3.9 Transformer Losses

The various losses occurring in a transformer are described below:

3.9.1 Iron or Core Losses (Magnetic Losses)

Iron loss occurs in the magnetic material of the transformer. This loss consists of: (i) hysteresis loss and (ii) eddy current loss. These two losses are given by empirical equations as

$$P_h = k_h f B_{max}^{1.6} v \text{ watts and } P_e = k_e (f B_{max})^2 v t^2 \text{ Watts} \tag{14}$$

where k_h = Steimetz hysteresis coefficient, depends on the quality of the material = 502.4 J/m^3 for good quality sheet steel and 191 J/m^3 for silicon steel

B_{max} = maximum flux density

v = volume of core, m^3

k_e = constant depending upon the electrical resistance of the magnetic material and grade of iron

t = thickness of the laminations of magnetic material

Hysteresis loss is a measure of the electric energy required to overcome the retentivity of the iron in the magnetic flux path. Eddy current loss is due the flow of eddy current in the core. This loss varies as the square of the thickness of laminations. In order to reduce this loss, thin laminations are used which are insulated from each other by a small layer of varnish.

The applied voltage V_1 can be written as

$$V_1 = E_1 = 4.44\Phi_{max} f T_1$$

$$\therefore B_{max} = \frac{V_1}{4.44 A_c f T_1} \tag{15}$$

$$\therefore P_h = k_h f \left(\frac{V_1}{4.44 A_c f T_1} \right)^{1.6} v \text{ and } \therefore P_e = k_e \left(\frac{V_1}{4.44 A_c T_1} \right)^2 v t^2 \quad (16)$$

The hysteresis loss is dependent on applied voltage and frequency. The eddy current loss is dependent on the square of the applied voltage, thickness of the lamination and independent of frequency.

Thus core losses are dependent on flux density, applied voltage, frequency and thickness of the laminations. For a particular transformer V_1, f and t are constant. The flux density varies with the change in load but the variation is negligible. Therefore the core losses are assumed to remain constant from no load to full load. These losses are determined from the ***open circuit test***.

3.9.2 Copper or Ohmic Losses (Electrical Losses)

The copper losses are the winding losses. The copper losses are present due to the resistance of the winding and the current flowing through the winding, i.e., I^2r losses. These losses vary as the square of the current. The total copper losses in the transformer are the addition of primary and secondary copper losses, i.e., $I_1^2r_1 + I_2^2r_2 = I_1^2R_1 = I_2^2R_2$, where R_1 and R_2 are resultant equivalent resistance referred to primary and secondary respectively. The copper losses are determined from the ***short circuit test.***

3.9.3 Stray Loss and Dielectric Loss

The stray loss and dielectric loss are usually small and therefore neglected. The stray load loss is due to the eddy current flow in the conductors, tanks. This eddy current is due to leakage flux.

The dielectric loss occurs in insulating materials, that is, transformer oil and solid insulation of the transformer.

3.10 Open Circuit Test

The purpose of this test is to determine the shunt branch parameters of the equivalent circuit of the transformer. As the name suggests the secondary HV side is open or on no load during this test. Since there is no load on secondary side, this test is also called ***no load test***. The primary LV is connected to supply and rated voltage is applied. It draws small amount of current (2 to 6%) from the supply known as no load current. This current is having two components as per equivalent circuit diagram shown in Fig. 3.1). One is to magnetize the core, known as ***magnetizing current*** or reactive component I_m or I_μ. Magnetizing current is in phase with Φ and is lagging by 90° to the applied voltage V_1. The other component is core loss or active component I_l or I_a and is in phase with the applied voltage V_1. The no load current I_0 lags behind the applied voltage V_1 by an angle θ_0 which is decided by the magnitude of its components I_m and I_l.

The power loss in the transformer is due to core loss and very small I^2R loss in the primary. There is no I^2R loss in the secondary since it is open and I_2=0.

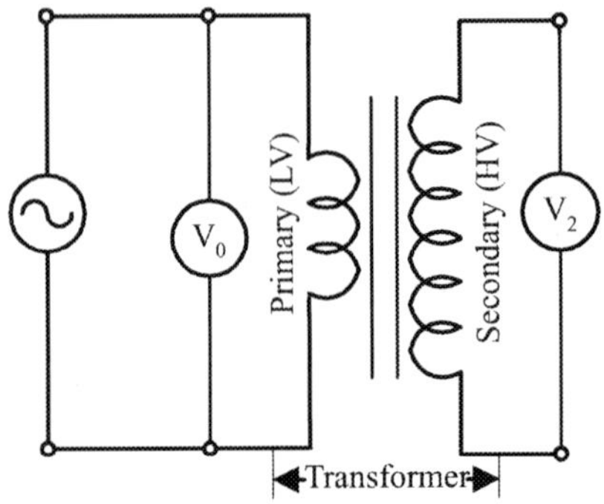

Fig. 3.11. Open circuited transformer.

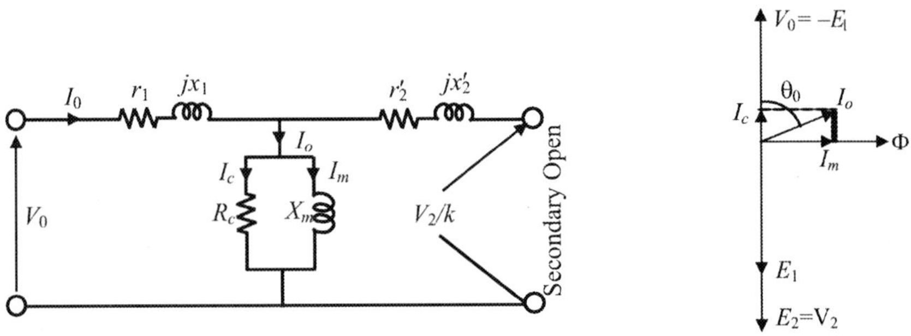

(a) Equivalent circuit diagram

(b) Phasor diagram

Fig. 3.12. Open circuit test.

Since no load current I_0 is very small, i.e., 2 to 6% of the full load primary current, the I^2R loss in the primary winding can be neglected. The core loss depends upon the flux. Since the rated voltage V_1 is applied, the flux set up by it will have a normal value so that normal core losses will occur. This core loss is the same at all loads. If wattmeter (low pf) is connected on input side, that will measure the core loss or iron loss only.

Measuring input power P_0, input voltage V_1 or V_0 (up to rated) and input current I_0 on the primary side, the active and reactive components of no load current can be separated and thus R_0 and X_m of the shunt branch of the equivalent circuit can be determined. The resistance r_1 can be measured by digital multimeter and that DC resistance can be converted to AC, $r_{ac} = (1.1 \textit{ to } 1.3)\, r_{dc}$. The value of R_c and X_m are very large as compared to r_1 and x_1. Therefore, equivalent circuit diagram consists of only R_0 and X_m in parallel.

Input power $P_0 = V_0 I_0 \cos \theta_0$; Therefore, $\cos \theta_0 = P_0 / (V_0 I_0)$; and

$I_l = I_c = I_0 \cos \theta_0$; $R_c = V_0/I_c$ ohms; $I_m = I_r = I_0 \sin \theta_0$; $X_m = V_0/I_m$ ohms

Generally, open circuit test is performed on low voltage side because on this side low voltage is required to be applied, which is available in the laboratory. On the high voltage side, the voltage can be measured by any high rating voltmeter. If the high voltage supply, and high rating meters are available in the laboratory, then with ***proper care*** of high voltage, this test can also be performed on high voltage side.

3.11 Short Circuit Test

The purpose of this test is to determine the series parameters of the transformer. When the transformer secondary terminals are short circuited, as shown in Fig. 3.13, the secondary current is large, because it is limited only by the transformer winding impedance. Neglecting the no load current, the short circuit current is given by

$$I_{sc} = \frac{\text{Voltage}}{\text{Transformer leakage impedance}}$$

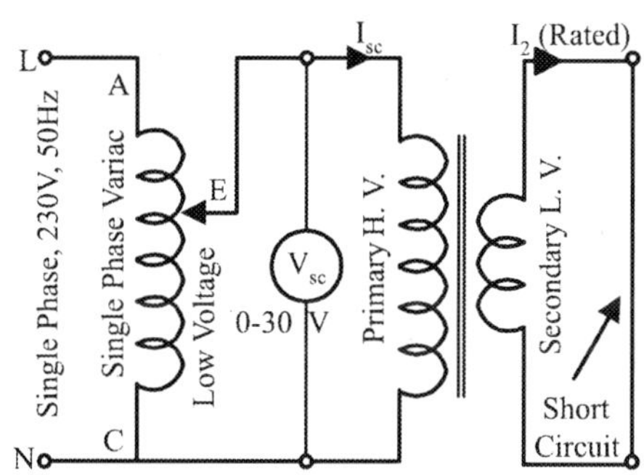

Fig. 3.13. Short circuit transformer.

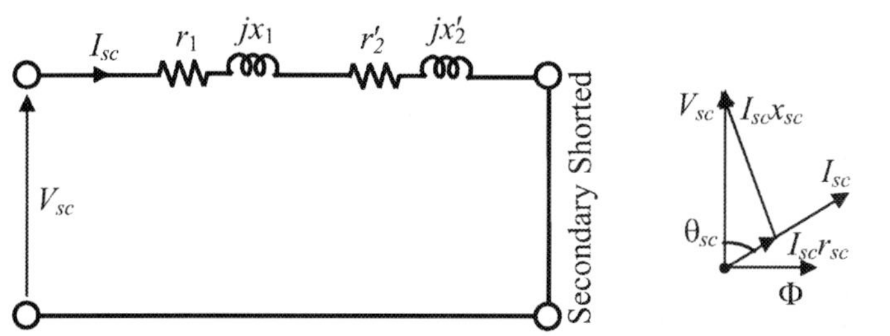

(a) Equivalent circuit diagram (b) Phasor diagram

Fig. 3.14. Short circuit test.

This current is large, since the transformer leakage impedance ($= r + jx_L$) is usually small. To keep this current within limit, the only way is to apply a ***low voltage*** to the transformer primary. Generally 4 to 8% of the rated voltage is required to circulate the rated current under short circuit condition. Since the voltage applied is low, the magnetizing current through the no load branch in its equivalent circuit (Fig. 3.14(a)), is low. Again the R_0 and X_m are very large as compared to the short circuit impedance ($r_{sc} + jx_{sc}$). Hence no load branch can be ignored. The equivalent circuit then becomes as shown in Fig. 3.14(a).

Usually, in the short circuit test, Fig. 3.13, the low voltage side is short circuited by a thick conductor. A wattmeter, a voltmeter and an ammeter are connected on high voltage side, i.e., the short circuit test is performed on high voltage side. The reasons for short circuiting the LV side and taking measurements on the HV side are:

1. The rated current on HV side is lower than that on LV side. This current can be safely measured with the available laboratory ammeters.
2. Since the applied voltage is very less (4 to 8%) than the rated voltage of the winding, greater accuracy in the reading of the voltmeter is possible when the HV side is used as the primary.

If the high current supply, and high rating meters are available in the laboratory, then with ***proper care***, this test can also be performed on low voltage side.

The supply voltage is gradually increased until full load primary current flows. When the rated full load current flows in the primary winding, rated full load current will flow in the secondary winding by transformer action. Since the applied voltage is low, the flux produced is low and the core loss which is nearly proportional to square of the flux, is so small and it can be neglected. However, the windings are carrying normal full load currents and therefore the input is supplying the normal full load copper losses. Thus the wattmeter gives the full load copper losses P_{sc}. The output voltage V_2 is zero because of short circuit. Consequently, whole of the primary voltage V_{sc} is used in supplying the voltage drop in the total impedance Z_{sc} referred to primary. If I_{sc} is the rated primary current then,

$$\text{Power factor } \cos\theta_{sc} = P_{sc} / (V_{sc} I_{sc}),\ Z_{sc} = V_{sc}/I_{sc};$$

$$(r_1 + r_2') = Z_{sc} \cos\theta_{sc},\ (x_1 + x_2') = Z_{sc} \sin\theta_{sc}.$$

When the primary resistance r_1 is measured, the secondary resistance, referred to primary r_2' can be separated from the total resistance, $r_2' = (r_1 + r_2') - r_1$.

The leakage reactance is generally assumed to be equal and hence, $x_1 = x_2' = (x_1 + x_2')/2$.

4. EXPERIMENTAL SET-UP

In order to perform the two tests, it is necessary to have two different set-ups.

4.(a) Open Circuit Test

4.(a).1 Apparatus Required

1. Single phase variac, single phase transformer, rating: for example: 2.5 kVA, 230/400 V.
2. Ammeter: One, 1/2 amps, or multimeter
3. Voltmeter: One, 0-300 volts. or clampmeter.
4. Wattmeter: One, 1/2 amps, 300 volts.

4.(a).2: Circuit Diagram

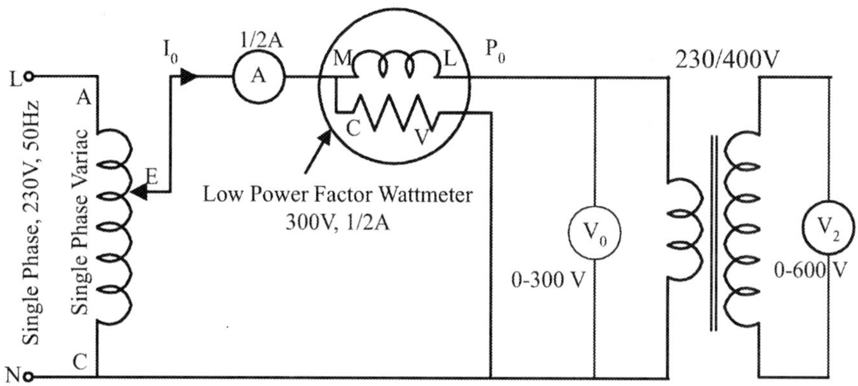

Fig. 3.15: Experimental set-up for open circuit test on a transformer.

4.(a).3 Procedure

1. Make the connections as shown in Fig. 3.15.
2. Set the variac to zero output and switch on the AC supply.
3. Increase the voltage by variac, up to the suitable value of voltage (say 200 V) (primary side).
4. Note down the meter readings.
5. Repeat step 3, up to rated voltage and repeat step 4.

Note: On no load, the transformer is operating on low power factor; therefore, for better results, low power factor wattmeter is used. The variac has been connected for safety purpose only. With the variac connected, the power input can be measured at different voltages and thus the variation in the core loss with voltage can be known.

4.(a).4 Observations

Sr. No.	Observations				Calculations		
	Primary side			Secondary side	Power factor	R_c	X_m
	Power P_0	Current I_0	Voltage V_0	Voltage V_2	$\cos\theta_0 = P_0/(V_0I_0)$	$= \dfrac{V_0}{I_0\cos\theta_0}$ Ohms	$= \dfrac{V_0}{I_0\sin\theta_0}$ Ohms
1.			200 V				
2.			210 V				
..			..				

4.(b) Short Circuit Test

4.(b).1 Apparatus Required

1. Single phase variac, single phase transformer, rating: 2.5 kVA, 230/400 V.
2. Ammeter: Two, 10 amps, and 20 amps or multimeter
3. Voltmeter: One, 0-15 volts, or clampmeter.
4. Wattmeter: One, 20 amps, 75 volts.

4.(b).2 Circuit Diagram

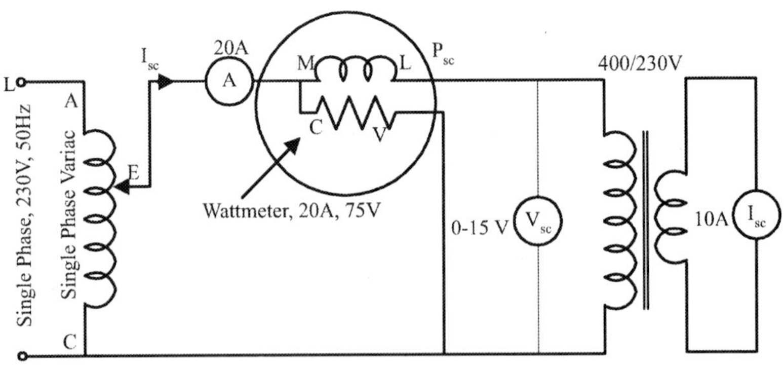

Fig. 3.16. Experimental set up for short circuit test on a transformer.

4.(b).3 Procedure

1. Make the connections as shown in Fig. 3.16.
2. Set the variac to zero output and switch on the AC supply.
3. ***Carefully apply low voltage*** by variac, up to the suitable value of current (say 9A) (primary side).

4. Note down the meter readings.
5. Repeat step 3, up to rated value of current (10.8A) and repeat step 4.

4.(b).4: Observations

Observations					Calculations			
Sr. No.	Primary Side			Secondary side voltage I_2	Power factor $\cos\theta_0 = P_{sc}/(V_{sc}I_{sc})$	$Z_{sc} = \frac{V_{sc}}{I_{sc}}$ ohms	$r_1 + r'_2 = Z_{sc} \cos\theta_{sc}$ ohms	$x_1 + x'_2 = z_{sc} \sin\theta_{sc}$
	Power P_{sc}	Current I_{sc}	Voltage V_{sc}					
1.		9 A						
2.		10 A						
3.		10.8 A						

5. RESULTS

5.(a) Open Circuit Test

The values of R_c and X_m are calculated and found out to be R_c =..... Ω and X_m =..... Ω and R_c =.....pu and X_m =..... pu
(pu = actual value / base value), base value of impedance = rated voltage / rated current)

5.(b): Short Circuit Test

The values of r_1, r_2', x_1 and x_2' are calculated and found out to be r_1 =..... Ω, r_2' = ... Ω, x_1 =..... Ω x_2' =.....Ω and r_1 =... pu, r_2' = ... pu, x_1 =..... pu x_2' =... pu
Draw the equivalent circuit diagram and write down the pu values of parameters in that diagram.

6. CONCLUSION

1. The open circuit test on a transformer gives the core losses. The copper loss is negligible due to small amount of no load current (2 to 6%).
2. The short circuit test on a transformer gives the copper losses. The core losses are negligible due to low value of input voltage (4 to 8%).

7. POST-EXPERIMENTAL QUIZ

1. Which loss do you get by performing the open circuit test?
2. In the open circuit test, which loss is negligible and why?
3. In the short circuit test, which loss is negligible and why?
4. Which precaution should be taken while performing the short circuit test?

CHAPTER 4

To Study: (a) the Polarity Markings on Single-Phase Transformer Windings and (b) the Operation of an Autotransformer

1. OBJECTIVES

The main objectives of this experiment are:

1. to mark the polarity on the windings of a transformer.
2. to study the sense of the winding with respect to induced emf.
3. to study the operation of an autotransformer.

2. PRE-EXPERIMENTAL QUIZ

1. What do you mean by sense of windings with respect to an induced emf?
2. What do you mean by polarity marking?
3. What are the bad effects of wrong polarity markings?

3. THEORY

3.1 Polarity Marking

The polarity of the voltage source is fixed. The positive (+) terminal is at higher potential and negative (–) potential is at lower potential. A voltage source of V volts is shown in Fig. 4.1.

When we move from the positive terminal to the negative terminal, there is a *voltage drop* and when we move from the negative terminal to the positive terminal, there is a *voltage rise*. If a resistor is connected across the voltage source, then

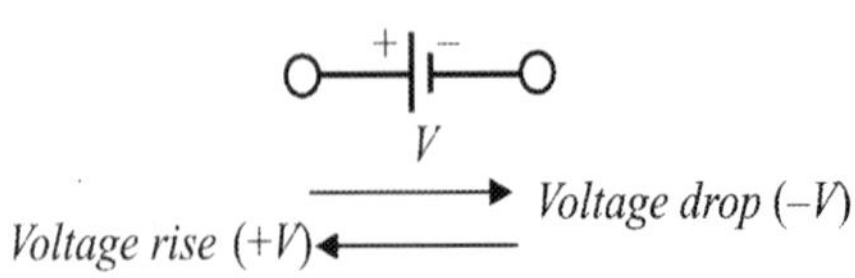

Fig. 4.1. Polarity of a voltage source.

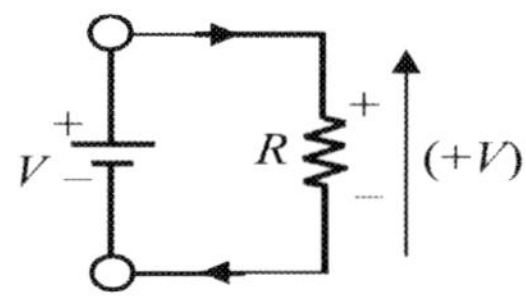

Fig. 4.2. Polarity of a voltage of a resistor.

current flows from higher potential point to lower potential point. In the direction of current there is a voltage drop and in opposite direction of current there is voltage rise. The positive (+) sign is assigned to the terminal of the resistor where the current enters and negative (–) sign is assigned to the terminal of the resistor where the current leaves it.

A change in current *i*, results into a change in Φ, which induces emf in the winding or coil wound on a magnetic core, is given by

$$v = -e = -L\frac{di}{dt} = -N\frac{d\Phi}{dt} \tag{1}$$

For coil 1, with number of turns N_1 the equation can be written as

$$v = -e_1 = -L_1\frac{di}{dt} = -N_1\frac{d\Phi}{dt} \tag{2}$$

If another coil with N_2 turns is put on the same core, the induced emf is given by

$$e_2 = L_2\frac{di}{dt} = N_2\frac{d\Phi}{dt} \tag{3}$$

The polarity of the induced emf depends on the sense of winding. For example consider three coils A, B and C, as shown in Fig.4.3.

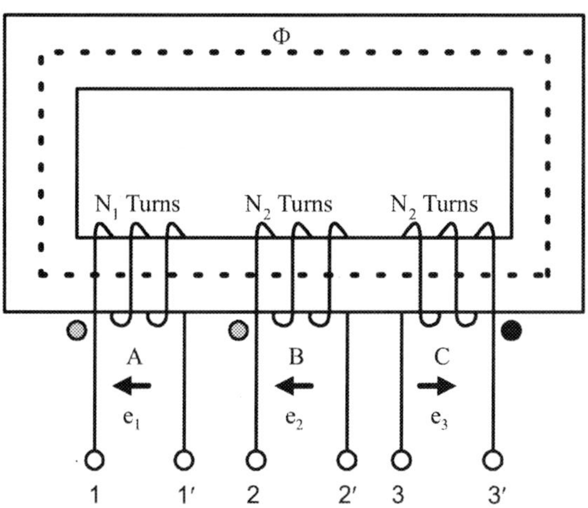

Fig. 4.3. The polarity of induced emfs for three coupled coils.

The coils *B* and *C* both have N_2 turns. As per the above equations, both will have the same magnitude of the induced emfs. Carefully note the sense of winding. Coils *A* and *B* have the same winding sense while coil *C* has the opposite winding sense. With the coil end marked as 1-1′, 2-2′, and 3-3′, if the current is entering at the ends 1, 2, and 3, then the coil mmfs are not all in one direction (Apply R.H.R.: if fingers wrap around the coil with the fingers pointed in the direction of current, the thumb of right hand indicates the direction of flux inside the coil). Such is the effect of sense of winding. The induced emfs are in one direction in the coils A and B, but in opposite direction in the coil C. This is as marked in Fig. 4.3. and induced emf e_1 is in coil *A*; e_2 in coil *B* and e_3 in coil *C*. However, those in B and C are equal but opposite. **They are equal due to the equal number of turns and are opposite due to different sense of windings.**

While writing down the equations for the electric circuit, the *SIGN* is most important. Hence, the polarity of the induced emf must be known. It is known by the *POLARITY MARKINGS* on the coils in Fig. 4.3, the identical coil ends are marked with identical dots • (■, ▲, ●, ☼, ۞ or any suitable marking).

The ends 1, 2 and 3' marked with dot. In other way, it also means that the undotted ends 1', 2' and 3 are identical. Such dot markings are essential in electric diagrams.

According to the dot convention, when the currents enter (or leave) the dotted ends, the mmfs and fluxes are additive.

3.1(a) Series Connection

If coils *A* and *B* are connected in series, a voltage *v* can be applied to the series combination and the same current will flow in the two coils. The coils have a mutual inductance of M_{ab} henrys. The current enters both the coils at their dotted end and hence identical ends. The mmfs must be additive and hence the fluxes must be *ADDITIVE*. The voltage induced in the two coils can be written as

$$\text{Total voltage induced in coil } A = e_a = L_a \frac{di}{dt} + M_{ab} \frac{di}{dt} \tag{4}$$

$$\text{Total voltage induced in coil } B = e_b = L_b \frac{di}{dt} + M_{ab} \frac{di}{dt} \tag{5}$$

Hence, supply voltage *v* is

$$v = e_a + e_b = \left(L_a + L_b + 2M_{ab}\right)\frac{di}{dt} \text{ (Additive Voltage)} \tag{6}$$

Now, if the coils *A* and *C* are connected in series in such a way that the current *i* enters the coil *A* at the dotted end, while for coil *C* the same current enters at the undotted end. The fluxes in the core will then be *OPPOSING* and hence subtractive. The following relations are then applicable.

$$\text{Total voltage induced in coil } A = e_a = L_a \frac{di}{dt} - M_{ac} \frac{di}{dt} \tag{7}$$

$$\text{Total voltage induced in coil } C = e_c = L_c \frac{di}{dt} - M_{ac} \frac{di}{dt} \tag{8}$$

Hence, supply voltage v is

$$v = e_a + e_c = \left(L_a + L_c - 2M_{ac}\right)\frac{di}{dt} \text{ (Subtractive voltage)} \tag{9}$$

In this case, the induced voltage is less. Therefore, more current from input may flow in the coils and sometimes coils may burn.

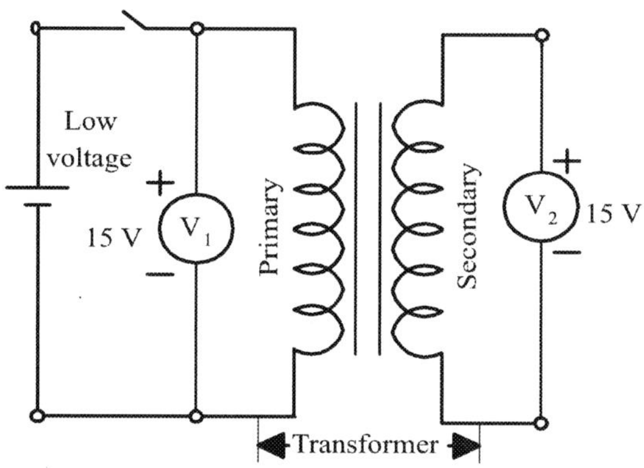

Fig. 4.4. Polarity marking with DC supply.

The polarity marking can also be done by using DC supply. Make the connection as shown in Fig. 4.4. Apply **low DC voltage**. Make the switch ON ***momentarily*** and watch the voltmeter V_2 (center tapped) on secondary side. If it deflects towards ***right***, i.e., towards positive, then the coil end connected to the positive terminal of the voltmeter V_2 is similar to the coil end (primary) which is connected to the positive of the source. If the voltmeter V_2 deflects towards left, i.e., towards negative, the conclusion is otherwise.

3.1(b) Three-Phase Transformer

In making three-phase transformer connections, proper attention must be given to the polarity markings. If the primary or secondary connections are done with wrong polarity markings, the transformer operation will be unsatisfactory and in some cases it may even burn. The importance of polarity markings can be appreciated by considering a transformer connected with wrong polarity markings.

Let a transformer with Yy0 connection be considered for this purpose. The connection diagram and the phasor diagram with proper polarity markings are

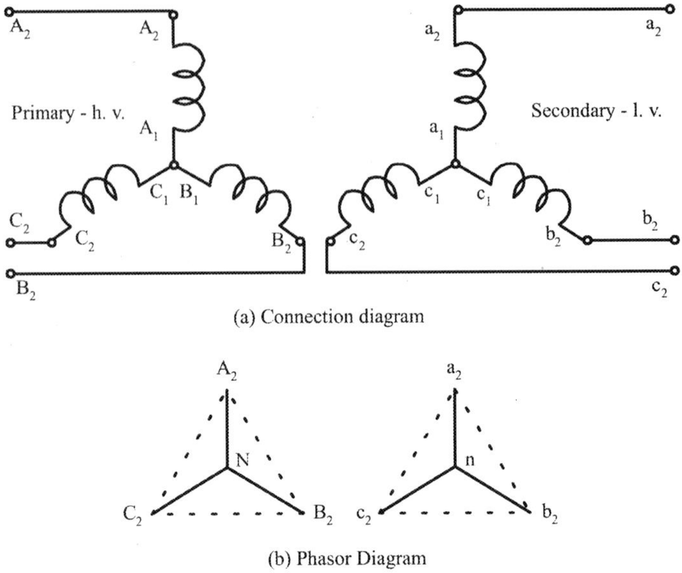

Fig. 4.5. Yy0 arrangement with correct polarity markings.

shown in Fig. 4.5. Line voltages are indicated by dotted lines joining A_2B_2, a_2b_2 etc., in Fig. 4.5. Now suppose the l. v. (secondary) winding connection a_1 and a_2 are reversed without altering h.v. (primary) winding connections A_1 and A_2 as shown in Fig. 4.6 (a). With this change l.v. phasor diagram of Fig. 4.5 (b) gets modified to that shown in Fig. 4.6 (b). Note that phasor a_2a_1 is reversed from the

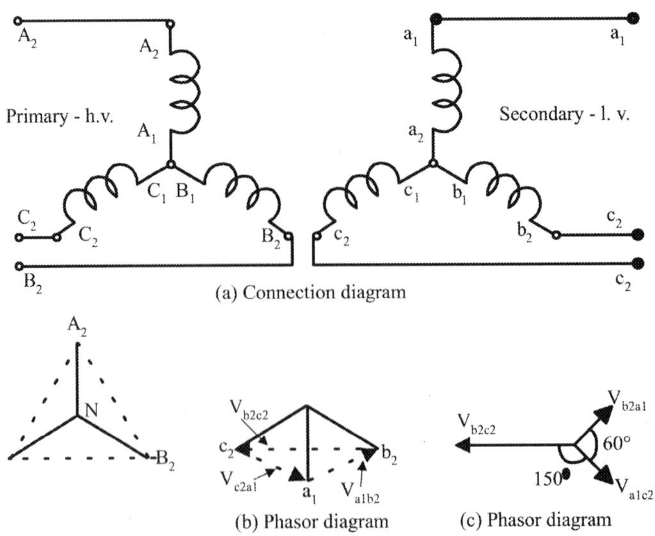

Fig. 4.6. Effect of connecting A-phase of secondary winding inaccurately.

position it occupied in Fig. 4.5 (b). The secondary line voltages are indicated by dotted lines in the phasor diagram of Fig. 4.6 (b). In order to show the line voltages clearly, the phasor diagram of Fig. 4.6 (c) is drawn. The three line voltages in l. v. sides are seen to be unbalanced. The line voltages Va_1b_2 and Vc_2a_1 have their magnitudes equal to the phase voltage and these are time displaced from each other by 60°. The line voltage V_{b2c2} is $\sqrt{3}$ times the phase voltage, but thus is time displaced by 150° from other two line voltages Va_1b_2 and Vc_2a_1. The same effect can be obtained if phase A terminals A_1A_2 of primary are reversed.

3.1.(c) Effects of Polarity Markings

1. There may be short circuit, flow of high current, which may result in burning of coil.
2. In case of three-phase line voltages decrease and become equal to phase voltage.
3. Three-phasors of line voltages are not displaced by an angle of 120°.
4. System becomes unbalanced.

3.2 Autotransformer

In a two-winding transformer, the primary winding is electrically isolated from the secondary winding. The two windings are coupled together magnetically by a common core. Thus, the principle of magnetic induction is responsible for the energy transfer from the primary to the secondary.

A transformer in which a part of winding is common to both primary and secondary circuit is called an autotransformer. Thus in an auto-transformer the two windings are not electrically isolated, but its theory and operation are similar to those of a two-winding transformer.

A simple arrangement of a step-down autotransformer is depicted in Fig. 4.7. N_1 and N_2 are the number of turns of winding AC (primary) and winding EC (secondary) respectively, input voltage is V_1, the exciting current is I_μ, flowing through the full winding AC and if the internal impedance drop is neglected, then the voltage per turn in winding AC is V_1/N_1 and the voltage across EC is (V_1/N_1) N_2. Therefore, output voltage, $V_2 = (V_1/N_1)\ N_2$. When the load is connected,

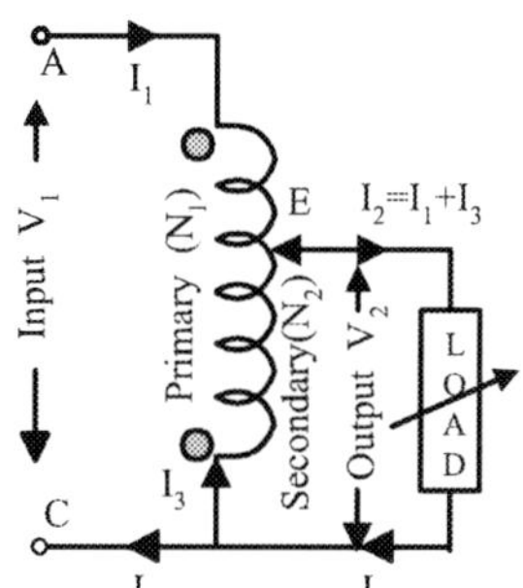

Fig. 4.7. Step-down auto-transformer.

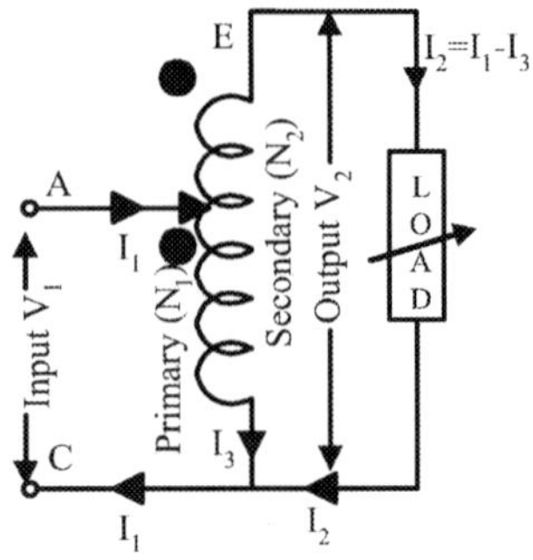

Fig. 4.8. Step-up auto-transformer.

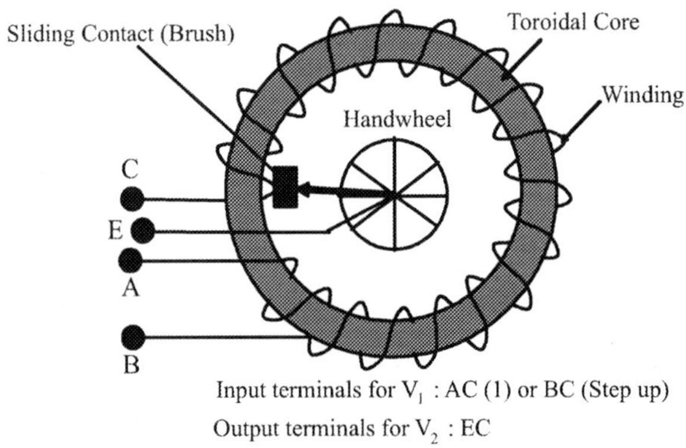

Fig. 4.9. Auto-transformer.

current I_2 starts flowing through the load and current I_1 is taken from the source. Neglecting losses and no load current, input power is equal to output power. Therefore, $V_1I_1 \cos \theta_1 = V_2I_2 \cos \theta_2$. If internal impedance drops and losses are neglected, then $\cos \theta_1 = \cos \theta_2$. Hence, $V_1I_1 = V_2I_2$, and $V_2/V_1 = I_1/I_2 = N_2/N_1 = k$.

On no load the exciting current flows from *A* to *C*, establishes a working mmf directed vertically downward in the core. When load is connected the current in winding *CE* must flow from *C* to *E*, in order to create an mmf opposing the exciting or working mmf as per Lenz's law. Since the working mmf in a transformer remains substantially constant at its no load value, the primary must draw additional current I_1 from the source, in order to neutralize the effect of current I_{ce} or I_3. In winding AC, I_1 flows from *A* to *C* while in winding *EC*, I_2 flows from *E* to *C*. For step down auto-transformer, since $V_2 < V_1$, $I_2 > I_1$, and therefore $I_2 = I_1 + I_3$.

A simple arrangement of a step-up auto-transformer is depicted in Fig. 4.8. For step up auto-transformer, since $V_2 > V_1$, $I_2 < I_1$, and therefore $I_2 = I_1 - I_3$.

3.2(a) Applications of Autotransformer

1. Auto-transformers are used for interconnecting the power systems of different voltages (132 kV and 220 kV).

2. Used as boosters to raise the voltage by small amount.
3. Autotransformers with a number of tappings are used for stating induction motors and synchronous motors. In this case these are called *auto-starters.*
4. Autotransformers are used to vary voltage continuously. In this case it is called *variac*, which is used in laboratory for testing.
5. Used in furnace for getting suitable supply to furnace winding.

3.2(b) Advantages of Autotransformer

1. Due to one winding, less winding material is required.
2. It is smaller in size and cheaper than the two winding transformer.
3. It has low ohmic and core loss and high efficiency as compared to two-winding transformer.
4. It has less leakage reactance and better voltage regulation.
5. The linearly varying output voltage can be obtained.

3.2(c) Disadvantages of Autotransformer

1. There is a direct contact between input and output. Under open circuit condition, full voltage would be applied to the load, which may damage the equipment.
2. Due to less internal impedance, large current flows under short circuit condition.

4. EXPERIMENTAL SET-UP

4.1 Polarity Marking

4.1 (a) Apparatus Required

1. Single phase variac, single phase transformer: Rating: For example, 2.5 kVA, 400/230 V.
2. AC voltmeter: three: 0-150 V or clampmeter.

4.1 (b) Circuit Diagram

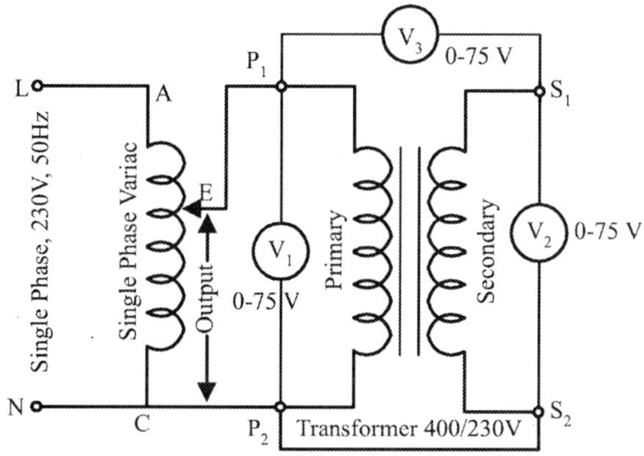

Fig. 4.10. Experimental set-up for polarity markings.

4.1(c) Procedure

1. Make the connection as shown in Fig. 4.10.
2. Apply single phase low voltage (say 30V) using single phase dimmerstat.
3. Read all the meters and note down all the readings in observation table.
4. Increase the voltage and repeat the steps 2 and 3.

4.1(d) Observations

Sr. No	V_1 Volts	V_2 Volts	V_3 Volts
.	.	.	.
..	..	..	..

4.1(e) Results and Conclusions

1. If $V_3 = V_1 + V_2$, then voltages are additive and dot markings are at P_1 and S_2.
2. If $V_3 = V_1 - V_2$, then voltages are subtractive and dot markings are at P_1 and S_1.

4.2 Autotransformer

4.2(a) Apparatus Required

1. Single phase variac, single phase transformer: Rating: For example, 2.5 kVA, 400/230 V.
2. Loading rheostat: 5 kW, 230 volts.
3. AC ammeter: Two: 10 amps, one: 2 amps. or multi-meter.
4. AC voltmeter: Three: 0-300 volts. or clamp meter.

4.2(b) Circuit Diagram

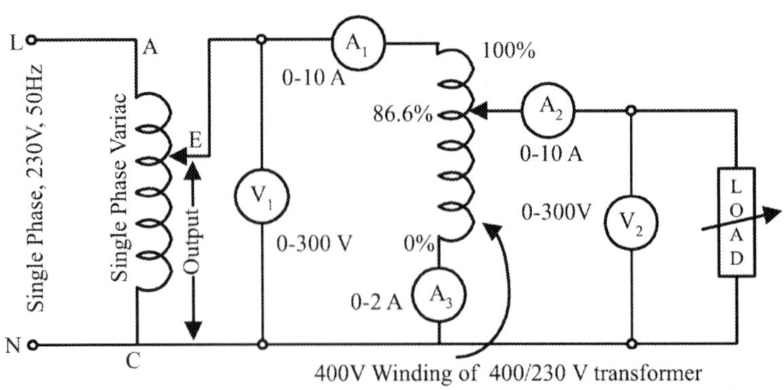

Fig. 4.11. Experimental set-up for studying a auto-transformer.

4.2(c) Procedure

1. Make the connection as shown in Fig. 4.11.
2. Apply single phase low voltage using single phase dimmerstat.
3. Switch on some load.
4. Read all the meters and note down all the readings in observation table.

4.2(d) Observations and Calculations

Sr.No.	A_1, I_1, Amp	A_1, I_2, A	A_3, I_3, A	V_1 Volts	V_2 Volts
.	.	.	.	.	.
..	..	..	..	..	..

1. $V_1I_1 = V_2I_2$
2. $I_2 = I_1 + I_3$ for step down autotransformer

4.3(e) Results and Conclusions

1. Verified that $V_1I_1 = V_2I_2$
2. Verified that $I_2 = I_1 + I_3$ for step down autotransformer.

5. POST-EXPERIMENTAL QUIZ

1. Two coils with opposite polarity markings are connected in series and then supply is given. Explain the result.
2. Explain the operation of and autotransformer.
3. What are the differences between a transformer and an autotransformer?
4. Explain the operation of a transformer when DC supply is given to the primary.

CHAPTER 5

To Study the Load Test on a Cumulatively Compounded DC Motor

1. OBJECTIVE

The main objective of this experiment is to study the effects of load on the efficiency and the output quantity, i.e., speed of the cumulatively compounded DC motor.

2. PRE-EXPERIMENTAL QUIZ

1. Write down the equation of speed for the cumulatively compounded DC motor.
2. What are the differences between cumulatively compounded and differentially compounded DC motor?
3. The conductors of series field winding have more diameter than shunt field winding. Why?

3. THEORY

3.1 Classification of DC Machines or Types or Methods of Excitation

In DC machine, basically there are two windings: (i) armature winding and (ii) field winding. The performance characteristics of a DC machine are greatly influenced by the way in which the field winding is excited with direct currents. There are in general two methods of exciting the field winding of DC machine: (i) separate excitation and (ii) self excitation.

3.1(a) Separate Excitation

Separately excited field winding consists of several hundred turns of fine wire or thin wire and is connected to a separate or external DC source as shown in Fig. 5.1. The voltage of the external DC source has no relation with the armature voltage. The field winding energized from a separate supply can be designed for

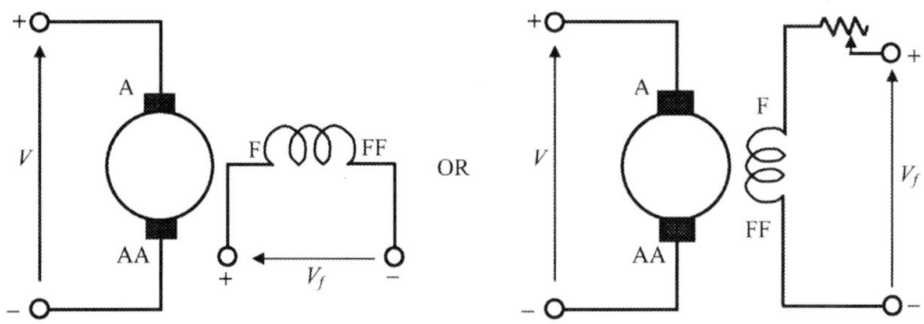

Fig. 5.1. Separately excited DC machine.

any convenient voltage. This machine is most flexible as full and independent control of both armature and field circuit is possible.

Permanent magnet machines also fall into this category. This motor consists of an armature and one or several permanent magnets encircling the armature. Schematically, this motor appears as an armature only.

Voltage equation of this motor is $V = E_b + I_a r_a$

Mechanical power developed is $P_m = VI_a - I_a^2 r_a = E_b I_a$

3.1(b) Self-excitation

When the field winding is excited by its own armature, the machine is said to be a self-excited DC machine. In these machines, the field poles must have a residual magnetism so that when the armature rotates, a residual voltage appears across the brushes. This residual voltage should establish a current in the field winding so as to reinforce the residual flux. A self-excited DC motor can be subdivided as: 1. series excitation 2. shunt excitation and 3. compound excitation.

1. Series Excitation or DC Series Motor or Series Wound DC Motor

The series field Y-YY winding consists of a few turns of ***thick*** wire with low resistance and is connected in series the armature as shown in Fig. 5.2. In other

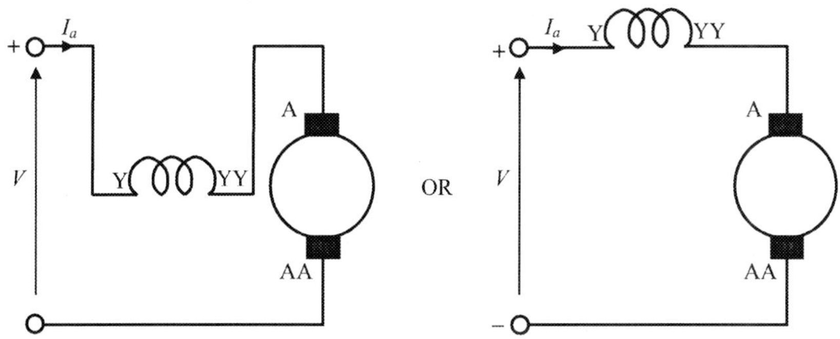

Fig. 5.2. DC series motor.

words the series field current depends on the armature current and in view of this a series field may be called *current operated field.*

Armature current, I_a = Series field current, I_{sc} = Line current, I_L

$$E_b = V - I_a(r_a + r_{sc})$$

$$P_m = V I_a - I_a^2(r_a + r_{sc}) = E_b I_a$$

2. Shunt Excitation or DC Shunt Motor or Shunt Wound DC Motor

The shunt field F-FF winding consists of a large number of turns of fine wire (so as to provide large resistance) and is connected in parallel or in shunt with the armature, as shown in Fig. 5.3. Therefore the voltage across the armature terminals and the shunt field is the same and it is for this reason a shunt field may be called a *voltage operated field.*

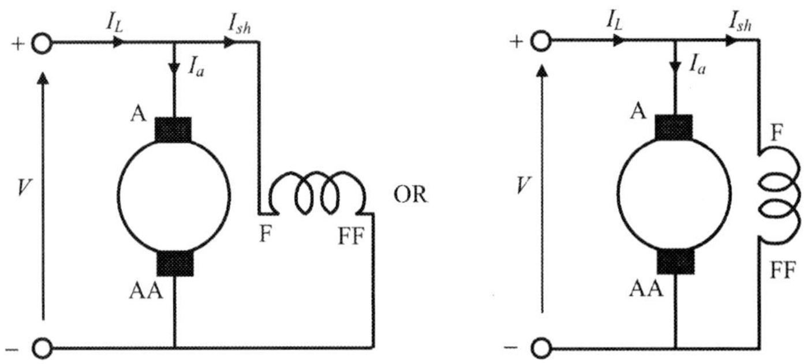

Fig. 5.3. DC shunt motor.

The series and shunt field are characterized by low and high resistances respectively. The current supplied to the motor is divided into two paths, one through the field winding and second through the armature, i.e., $I_L = I_a + I_{sh}$ and $I_{sh} = V/r_{sh}$

Back emf, $E_b = V - I_a r_a$ and power drawn by supply main, $P_i = V I_L$

$$P_m = V I_L - I_a^2 r_a - I_{sh}^2 r_{sh} = V I_L - I_a^2 r_a - V I_{sh}$$

$$= V (I_L - I_{sh}) - I_a^2 r_a = V I_a - I_a^2 r_a = I_a (V - I_a r_a)$$

$$= E_b I_a$$

3. Compound Excitation or Compound Wound DC Motor

Figure 5.4 shows how the shunt and series field windings are arranged on one pole. In the Fig. 5.4 (a) shunt field coil is placed near to yoke and series field coil is placed near the pole shoe just for the sake of clarity. Figure 5.4 shows the actual physical arrangement of these coils. It is seen that the first shunt field coil is

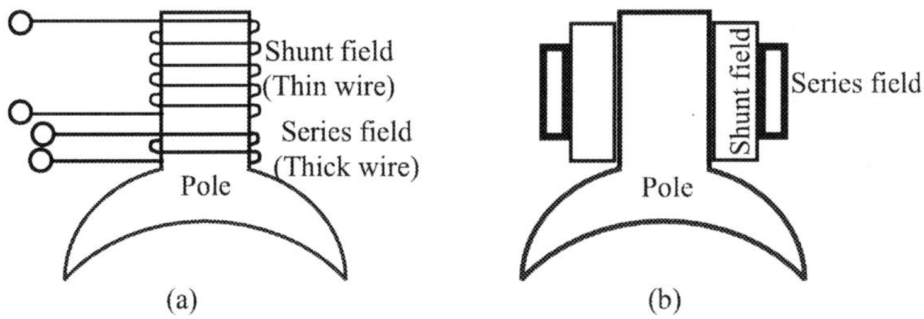

Fig. 5.4. Arrangement of shunt and series field windings on one pole.

wound around the pole body and over it series field coil is wound. The arrangement is for convenient in construction and for better cooling.

A DC compound excitation involves both the excitations, one series field Y-YY excitation and second shunt field F-FF excitation. A **shunt** field winding is connected in shunt or in parallel with the armature winding. This winding is the **main** winding or strong winding from the point of view of establishing flux. This winding has more number of turns of **thin** wire. In some applications, a shunt excitation winding may be replaced by a separately excited winding.

A **series** field winding is connected in series with the armature winding. In order to carry rated load current, this winding has less number of turns of **thick** wire. This winding helps to increase or decrease the flux. Depending on the connection of series field winding, compound wound motors are of two types: (i) cumulatively compounded DC motor and (ii) differentially compounded DC motor.

(1) Cumulatively Compounded DC Motor

If the series field winding is connected in such a way that the total air gap flux per pole increases or if the total flux is $\Phi_T = \Phi_f + \Phi_y$, then the motor is called cumulatively compounded DC motor. The series field strengthens the shunt field. The circuit diagram is as shown in Fig. 5.5.

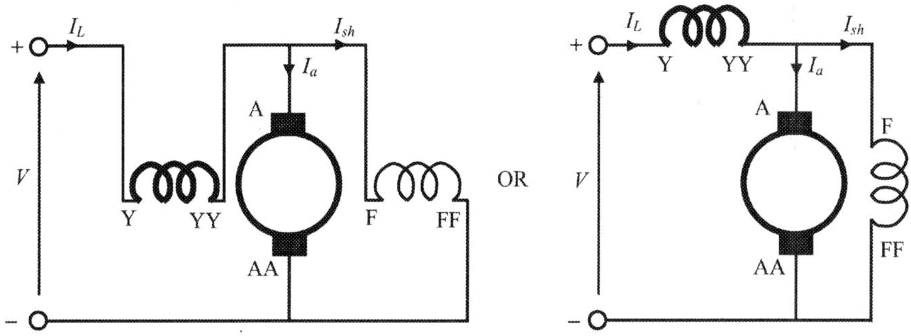

Fig. 5.5. Cumulatively compounded DC shunt motor.

(2) Differentially Compounded DC Motor

If the series field winding is connected in such a way that the total air gap flux per

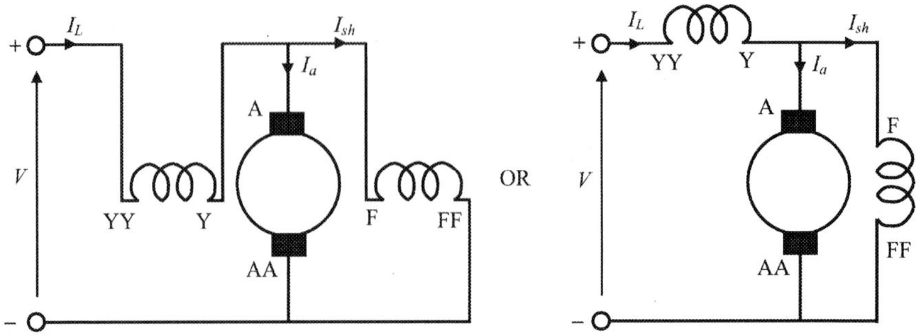

Fig. 5.6. Differentially compounded DC shunt motor.

pole decreases or if the total flux $\Phi_T = \Phi_f - \Phi_y$, then the motor is called differentially compounded DC motor. The series field weakens the shunt field. The circuit diagram is as shown in Fig. 5.6.

From the viewpoint of connection, DC compound motor has short-shunt connection and long-shunt connection as shown in Fig. 5.7. In short-shunt connection, the shunt field is connected across the armature terminals. The armature and shunt field are in parallel and this pair is in series with the series field, as shown in Fig. 5.7(a). In long shunt connection, the shunt field is connected across the line terminals, as shown in Fig. 5.7(b). The series field and armature are in series.

Back emf, $E_b = V - I_a r_a - I_L r_{se}$ and power drawn by supply main, $P_i = V I_L$

$$P_m = V I_L - I_a^2 r_a - I_{sh}^2 r_{sh} - I_L^2 r_{se} = V I_L - I_a^2 r_a - V I_{sh} - I_L^2 r_{se}$$

(Neglecting voltage drop across series field winding)

$$= V(I_L - I_{sh}) - I_a^2 r_a - I_L^2 r_{se} = V I_a - I_a^2 r_a = I_a (V - I_a r_a - I_L r_{se}) \text{ (Since } I_L \approx I_a)$$

$$= E_b I_a$$

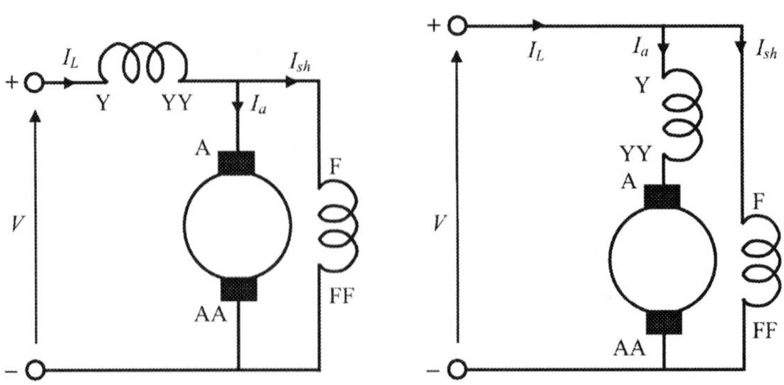

Fig. 5.7. Cumulatively compounded DC shunt motor.

3.2 Operating Characteristics

The performance and suitability of a DC motor is determined from its characteristics, known as *operating characteristics*. There are three important characteristics of DC motor.

1. Speed –Armature Characteristics

This characteristic gives the relation between speed N and armature current I_a. This is also known as *speed characteristic.*

(a) DC Shunt Motor and Separately Excited DC Motor

In both cases the field is supplied from a constant voltage so that the field current is constant and the air gap flux remains constant. The speed of the DC motor is given by the relation

$$N = K_n \frac{V - I_a r_a}{\Phi} \text{ rpm} \tag{1}$$

where Φ = field flux, in weber per pole.

For DC shunt motor, the speed is directly proportional to armature voltage and inversely proportional to field flux. For constant supply voltage and constant field current the speed is affected by $I_a r_a$ drop. With increase in load, I_a increases and speed decreases. But the resistance r_a is very small and negligible, so the voltage drop $I_a r_a$ is negligible. Then the speed (no load) equation becomes

$$N_0 = K_N \frac{V}{\Phi} \text{ rpm} \tag{2}$$

Thus, the speed of DC shunt motor remains constant. If resistance r_a is considered, then the speed drop is 6 to 8%. The speed armature current characteristics, is shown in Fig. 5.8.

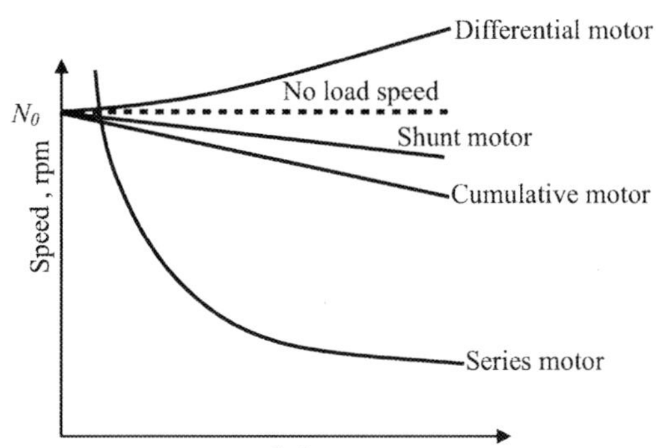

Fig. 5.8. Speed vs armature current, I_a characteristics.

(b) DC Series Motor

In DC series motor, since the field winding is connected in series with the armature, the current flowing through the field winding is same as the armature current, $I_a = I_y = I_{se}$. If r_{se} is the resistance of series field then the speed is given by

$$N = K_N \frac{V - I_a(r_a + r_{se})}{\Phi} \text{rpm} \tag{3}$$

$$N = K_{NSE} \frac{V - I_a(r_a + r_{se})}{I_a} \text{rpm} \tag{4}$$

If resistances are neglected, then the equation becomes

$$N = K_{NSE} \frac{V}{I_a} \text{rpm} \tag{5}$$

Thus the speed of series motor is inversely proportional to armature current and speed current characteristic is a *hyperbola*, as shown in Fig. 5.8. At no load, the armature current is very small. Due to this the speed becomes *dangerously high.* This will destroy the armature. In view of this, a series motor should ***never*** be started on no load, even accidentally. The minimum load should not be less than 15% of full load.

(c) Cumulatively Compounded DC Motor

A compound motor has both shunt and series field windings, so its characteristics are intermediate between shunt and series motors. The cumulative compound motor is generally used in practice. For cumulatively compounded DC motor, the speed is given by

$$N = K_N \frac{V - I_a(r_a + r_{se})}{\Phi_f + \Phi_y} \text{rpm} \tag{6}$$

The no load speed is determined mainly by shunt winding since the series field contributes little to the total flux under no load condition. Therefore, the no load speed is

$$N_0 = K_N \frac{V}{\Phi_f} \text{rpm} \tag{7}$$

Thus, the no load speed of cumulatively compounded DC motor is same as that of DC shunt motor. As the load increases I_a and Φ_y increase, total flux increases, speed decreases more than DC shunt motor. The speed armature current characteristic is shown in Fig. 5.8 where the characteristic is shown below the characteristic of DC shunt motor. By varying the proportions of shunt field and series field mmf, this motor may be designed to give the characteristic anywhere intermediate between those of shunt and series motors. If series winding is shorted, this can be operated as shunt motor and if shunt winding is open circuited, this can be operated as series motor. The cumulative compound motor has the following advantages:

1. At no load, like a shunt motor it can be safely started.
2. It has a high starting torque at no load speed.
3. It has more overload sustaining capacity than a shunt motor.

These factors make it suitable for use with heavy intermittent or fluctuating loads such as lifts, hoists, etc.

(d) Differentially Compounded DC Motor

In this case, the series field flux opposes the shunt field flux, therefore the speed is given by

$$N = K_N \times \frac{V - I_a(r_a + r_{se})}{\Phi_f - \Phi_y} \text{ rpm} \tag{8}$$

No load speed is

$$N_0 = K_N \times \frac{V}{\Phi_f} \text{ rpm} \tag{9}$$

As the load increases I_a and Φ_y increase, total flux decreases, speed increases. Since $I_a(r_a + r_{se})$ is small, can be neglected. So the speed increases with load. Due to this, motor will destroy itself. Because of this speed instability, this motor is no longer used in practice. The speed armature current characteristic is shown in Fig. 5.8.

2. Torque –Armature Characteristics

This characteristic gives the relation between torque developed in the armature, T_e and armature current. This is also known as *electrical characteristic.*

(a) DC Shunt Motor and Separately Excited DC Motor

The torque of the DC motor is given by the expression

$$T_e = K_T \Phi I_a \text{ Nm} \tag{10}$$

For constant flux, the torque increases linearly with the increase in armature current. Hence torque – armature current characteristic is a *straight line characteristic,* as shown in Fig. 5.9. The useful torque or shaft torque is, of course, less than the total torque or gross torque. This is due to the lost torque in iron, friction and windage losses.

(b) DC Series Motor

The torque of the DC series motor is given by the expression

$$T_e = K_I I_a^2 \text{ Nm} \tag{11}$$

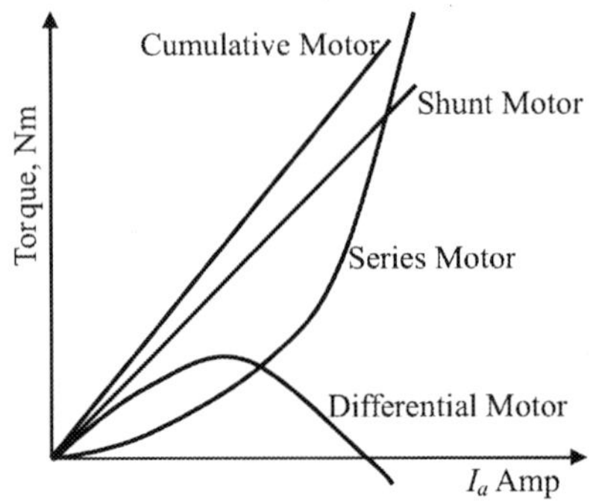

Fig. 5.9. Torque vs armature current, I_a characteristics.

Thus the torque is directly proportional to the square of the armature current, I_a and therefore, the torque-armature current characteristic is *parabolic in nature*, as shown in Fig. 5.9. It is evident that the starting torque is very high which is useful to accelerate the heavy masses like an electric train.

(c) Cumulatively Compounded DC Motor

The electromagnetic torque of the cumulatively compounded DC motor is

$$T_e = K_T (\Phi_f + \Phi_{se}) I_a \text{ Nm} \tag{12}$$

Thus the torque is directly proportional to the armature current, I_a. As the armature current increases, series field flux Φ_{se}, increases. As a result torque also increases. Thus the torque developed is higher than that of DC shunt motor, as shown in Fig. 5.9.

(d) Differentially Compounded DC Motor

The electromagnetic torque of the differentially compounded DC motor is

$$T_e = K_T (\Phi_f - \Phi_{se}) I_a \text{ Nm} \tag{13}$$

Initially as the motor is started that runs in one direction, say anticlockwise. Now with the increase in load, the armature current increases, series field flux Φ_{se} increases, as a result torque decreases. For more load, series field current and flux dominates the shunt field. The torque gets reversed and motor starts running on opposite direction, i.e., in clockwise direction. When the motor passes through zero speed, armature current, $I_a = V/r_a$, becomes dangerously high and the commutator and armature winding may get damaged.

Thus, due to high armature current at the time of starting and under more load condition, motor possesses inherent instability. Therefore, this motor is rarely used in industry. The torque armature current characteristic is shown in Fig. 5.9.

3. Speed-Torque Characteristics

This characteristic gives the relation between speed and torque developed in the armature, T_e. This is also known as *mechanical characteristic*. Since the torque is directly proportional to armature current, the speed torque characteristics have the same nature as that of speed armature current characteristics.

3.3 How to Start and Stop the Motor

At the time of start, the speed $N = 0$, so $E_b = 0$ and the armature current $I_a = (V - E_b)/(r_a + r_{se}) = V/(r_a + r_{se})$ which is very high and more than rated armature current. Therefore additional resistance is required to limit the current. This additional resistance is inserted in between line and armature by the use of starter. Hence, STARTER is needed to start the DC compound motor. During start, move the handle of the starter *SLOWLY* from off to on. At on position, the handle is held up by No Volt release coil. For stopping the motor, simply *DISCONNECT* the supply. Do not TOUCH the handle. After disconnecting the supply, handle will come automatically to its OFF position by spring action.

4. APPARATUS REQUIRED

1. DC Compound Motor: Rating: (For example 5kW, 230 V. 1000 rpm) Coupled DC Generator (Shunt type): Rating: (For example 5 kW, 230 V, 1000 rpm).
2. DC Ammeter: Two, 0-25 amps.
3. DC Voltmeter: Two, 0-300 volts.
4. Rheostats: Two, 370 ohms, 2 amps.
5. Tachometer or speedometer: one.
6. Loading Rheostat: 5 kW, 230 V.

5. CIRCUIT DIAGRAM

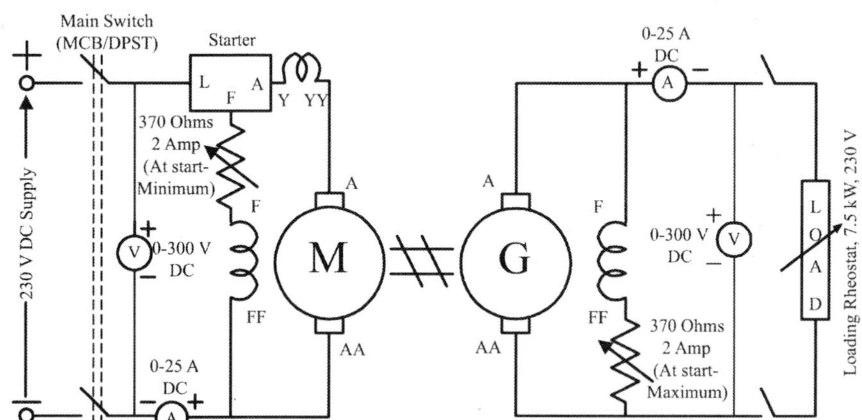

Fig. 5.10. Experimental set-up for load test on cumulatively compounded DC motor.

6. PROCEDURE

1. Make the connection as shown in Fig. 5.10.
2. Set up motor rheostat to minimum value and the generator field rheostat to maximum value.
3. Switch on the DC supply with the help of main switch or MCB. Move the handle of starter slowly and start the DC motor.
4. Adjust the speed of motor to a rated speed of generator with the help of motor rheostat.
5. Decrese the resistance of field rheostat of generator to build up the rated terminal voltage of generator. This voltage should be kept at this value (i.e., constant) for all the readings as far as possible. This is done in order to see the effect on the speed of the motor.
6. Load the generator with the help of loading rheostat and take down the reading of the voltmeters, ammeters and speed. As far as possible, take the readings trying to cover the range from no load to full load of the motor.

7. OBSERVATIONS

Observations						Calculations			
Sr. No.	Motor side			Generator side		Gen output	Motor output	Motor input	Motor efficiency
	Voltage V_m	Current I_m	Speed r.p.m.	Voltage V_g	Current I_g	V_gI_g	$=\frac{V_gI_g}{0.85}$	V_mI_m	$=\frac{Vg\,Ig}{0.85V_mI_m}$
1.									
2.									
..				Constant					
..									

8. RESULTS

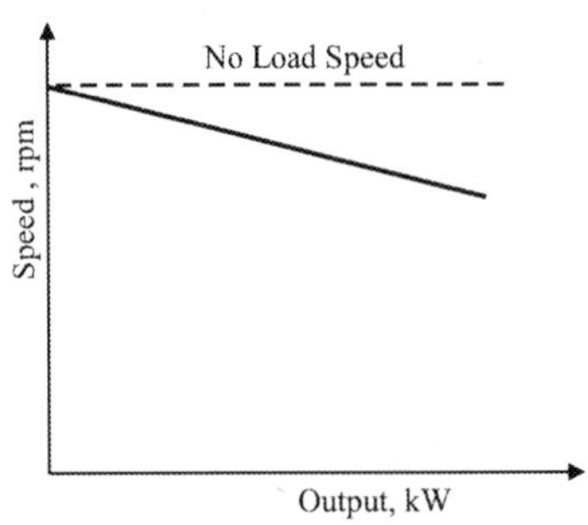

Fig. 5.11. Speed vs output characteristic.

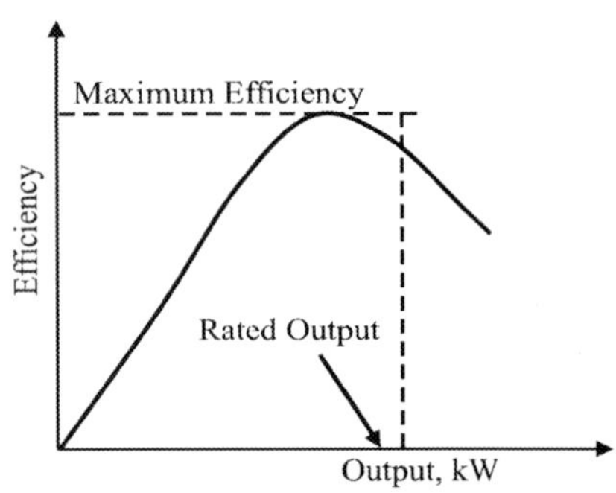

Fig. 5.12. Efficiency vs output characteristic.

The speed and efficiency vs output of motor has been plotted and studied. The characteristics are shown in Figs. 5.10 and 5.11 respectively.

9. CONCLUSION

1. The speed of cumulatively compounded DC motor decreases as the load on motor increases. The decrease in speed is more than DC shunt motor due to addition of flux.
2. The efficiency is zero at no load. It increases as the load increases and is maximum when variable losses are equal to constant losses.
3. The efficiency at rated output is less than the maximum value and the operating point is after maximum efficiency point.

10. POST-EXPERIMENTAL QUIZ

1. Explain the speed characteristics of DC shunt and DC cumulatively compounded motor, under load condition?
2. The no load speed of cumulatively compounded DC motor has finite value. Why?
3. DC cumulatively compounded motor is suitable for intermittent load. Why?

CHAPTER 6

To Study the Load Test on a Three-Phase Induction Motor

1. OBJECTIVE

The main objective of this experiment is to study the effects of load on the efficiency and the output quantity, i.e., speed of the three-phase induction motor.

2. PRE-EXPERIMENTAL QUIZ

1. Why a three-phase induction motor is self starting motor?
2. Write down the types of three-phase induction motor.
3. How the starting torque of a three-phase induction motor is increased?

3. THEORY

In 1892 at the Frankfurt exhibition Nikola Tesla, for the first time, presented a crude type of polyphase induction motor. Since then great improvements have been made in the design and performance of the induction motor. Due to its simple, rugged, reliable, efficient, inexpensive construction and excellent operating characteristics, these motors are very popular in industrial applications. As a rough estimate nearly 80% of the world's AC motors are induction motors. Due to the improvement in power electronics, nowadays, three-phase induction motors are used in electric locomotives.

It has good speed regulation and high starting torque. It requires little maintenance. It has a reasonable overload capacity.

The basic principle of operation is '*induction*' and hence the name '*induction motor*'. Induction is a phenomenon of an induced voltage in a coil due to changing flux established by either another coil (as a general case) or the same coil.

3.1 Construction

Similar to other rotating machines, an induction motor essentially consists of basic two parts: the stator and the rotor. The stator is the stationary part and the rotor is

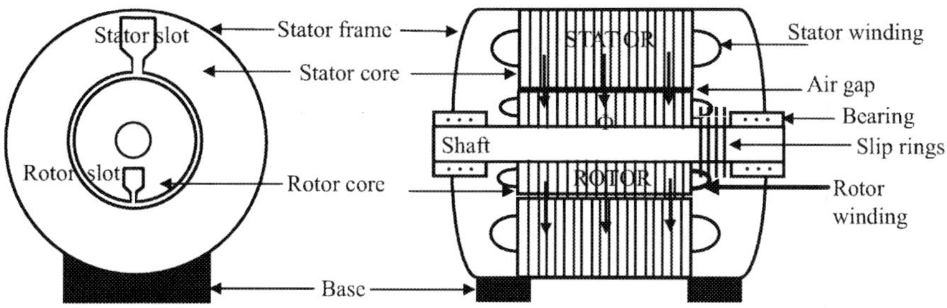

Fig. 6.1. Constructional features of rotating electrical machine.

the rotating part. The stator consists of stator frame, stator core, distributed winding, two end covers, bearing, etc. The stator core is made of high-grade alloy steel laminations (cold rolled grain oriented silicon steel sheets) to reduce eddy current losses. The laminations are *slotted* on the inner periphery and are insulated from each other. These laminations are supported in a stator frame of cast-iron or fabricated steel plate. The stator frame provides only mechanical support to the stator core and is not designed to carry the stator flux. The insulated stator conductors are placed in these slots. The stator conductors are placed in these slots. The stator conductors are connected to form a three-phase winding. The phase winding may be either star or delta connected. The end covers are made up of cast iron.

The clearance between the stator and the rotor is called an *air gap*. The air gap is kept as small as possible to reduce leakage reactance, no load current and to improve the power factor.

The rotor is built up of thin laminations of the same material as stator. Because of the lower frequencies of the rotor flux, thicker laminations can be used without excessive core loss. The laminated cylindrical core is mounted directly on the shaft or a spider carried by the shaft. The laminations are slotted on their outer periphery to house the rotor conductors or rotor winding. In order to provide cooling to the machine, fan blades are generally mounted on the rotor core which circulates the cooled air. There are two types of induction motor depending on the type of rotor.

1. Squirrel cage rotor or simply cage rotor (SCIM).
2. Phase wound or wound rotor or slip ring rotor (SRIM).

1. Squirrel-Cage Induction Motor

It consists of a cylindrical laminated core with slots nearly parallel to the shaft axis or *skewed*. Each slot contains an uninsulated bar conductor of aluminium or copper. At each end of the rotor, the rotor bar conductors are short circuited by a heavy end ring of the same material. For good electrical connections, the bars are riveted, brazed or welded with the two end rings. The conductors and the end rings look like a 'cage' of a bird or form a cage of the type which was once commonly used

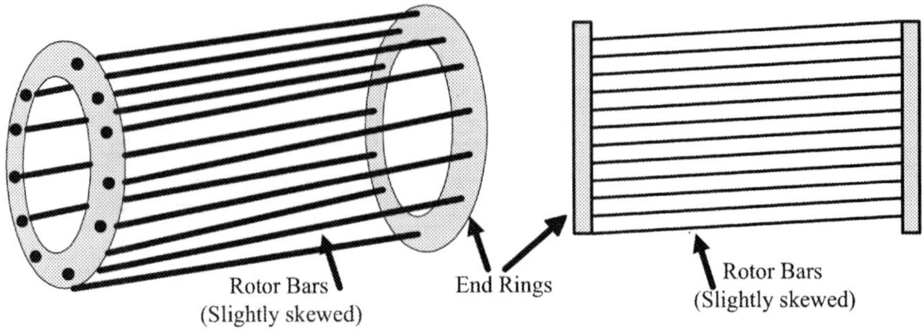

Fig. 6.2. Cage rotor.

for keeping squirrel; hence this rotor is known as the squirrel cage rotor. The important features of this rotor are:

1. The skewed bar gives uniform torque and less noise. The locking tendency is reduced.
2. It has a compact and rugged construction.
3. The end rings can be projected for fanning action for cooling.
4. It requires no slip rings.
5. It has less loss and more efficiency as compared to slip ring induction motor.
6. It is not possible to add extra rotor resistance to change the torque and speed. This is the only disadvantage.

2. Phase Wound or Wound Rotor or Slip Ring Rotor (SRIM)

The wound rotor consists of a slotted armature. Insulated conductors are put in the slots and connected to form a three-phase distributed double layer winding similar to the stator winding. The rotor windings are connected in star or in delta. The three ends of rotor windings are brought outside the rotor and connected to three insulated slip rings. The slip rings are mounted on the shaft with brushes resting on them. The brushes are connected to three variable resistors connected in star. The purpose of slip rings and brushes is to provide a means for connecting external resistors in the rotor circuit. The resistors enable to increase each rotor phase resistance to serve the following purposes:

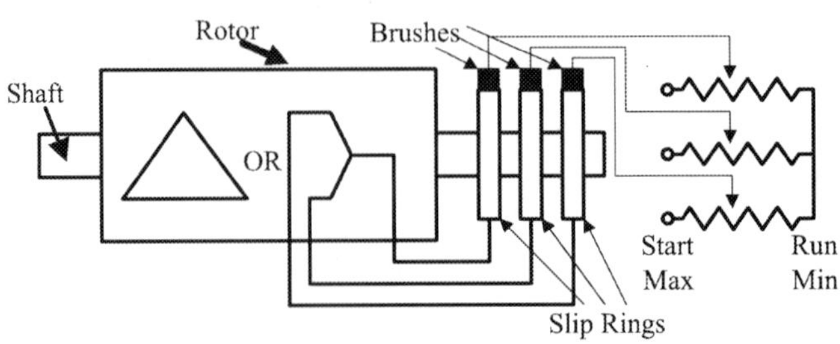

Fig. 6.3. Slip ring rotor.

1. to increase the starting torque
2. to decrease the starting current
3. to improve the starting power factor
4. to decrease the speed of the motor

Since the rotor is wound with polyphase windings and carries slip rings, it is called *wound rotor* or *slip ring induction motor.*

3.2 Principle of Operation

For the sake of simplicity, let us consider one conductor on the stationary rotor as shown in Fig. 6.4(a). Let this conductor be subjected to the rotating magnetic field produced when a three-phase supply is connected to the three-phase winding of the stator. Let the rotation of the magnetic field be clockwise. A magnetic field moving clockwise has the same effect as a conductor moving anticlockwise in a stationary field. By Faraday's law of electromagnetic induction, a voltage will be induced in the conductor. Since the rotor is complete, either through the end rings or an external resistance, the induced voltage causes a current to flow in the rotor conductor. By the right-hand rule we can determine the direction of induced current in the conductor. By this rule the direction of the induced current is outwards, represented by a dot, as shown in Fig. 6.4(b). The current in the rotor conductor produces its own magnetic field, as shown in Fig. 6.4(c).

We know that when a conductor carrying current is put in a magnetic field a force is produced on it. Thus a force is produced on the rotor conductor. The

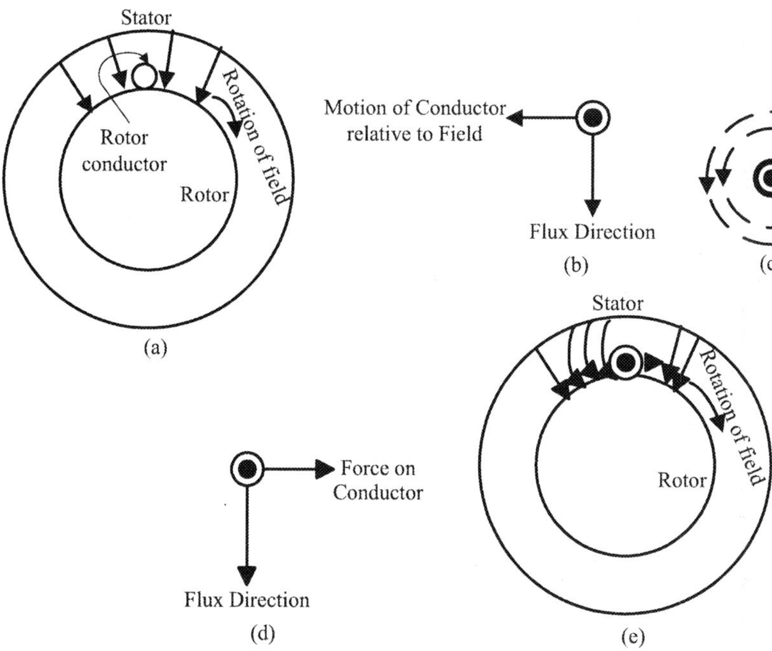

Fig. 6.4. Production of torque.

direction of this force can be found by the left-hand rule, as shown in Fig.6.4(d). It is seen that the force acting on the conductor is in the same direction as the direction of the rotating magnetic field. Since the rotor conductor is in a slot on the circumference of the rotor, this force acts in a tangential direction to the rotor and develops a torque on the rotor. Similar torques are produced on all the rotor conductors. Since the rotor is free to move, it starts rotating in the *same direction as the rotating magnetic field*. Thus a three-phase induction motor is self starting. Since the operation of this motor depends on the induced voltage in its rotor conductors, it is called an *induction motor*.

With the application of Lenz's law also the direction of rotation of the rotor can be determined. The induction of voltage and flow of current in the rotor conductors is due to relative speed between the rotor and the rotating magnetic field. According to this law, this current should oppose the cause of it and cause is relative speed. Therefore, in order to decrease the relative speed, rotor has to rotate in the same direction as that of rotating magnetic field.

An induction motor ***cannot*** run at synchronous speed. Let us consider that the rotor is rotating at synchronous speed. Under this condition, there would no cutting of flux by the rotor conductors, and there would be no generated voltage, no current and no torque. Therefore the rotor speed is slightly less than the synchronous speed. As soon as the speed decreases, due to relative speed between rotor and rotating magnetic field, rotor conductors cut the flux, rotor voltage is induced, rotor current flows and torque is developed. In this case the rotor inertia is the load on the motor. So induction motor runs at a speed depending on this inertia load. This speed is less than synchronous speed. Therefore an induction motor may also be called *asynchronous motor.*

If primary three-phase winding is on rotor and three-phase supply is given to this winding, then rotor has to rotate in opposite direction to that of rotating magnetic field in order to reduce the relative speed between rotor and rotating magnetic field.

3.3 Slip

The difference between the synchronous speed and the actual rotor speed is called the *slip speed*. The ***slip*** is defined as the ratio of slip speed to the synchronous speed. The slip at full load varies from 2 to 5%. By definition, the slip can be written as

$$s = \frac{N_s - N_r}{N_s} \tag{1}$$

The relative speed between stator flux and rotor conductor is $N_s - N_r$, which is equal to sN_s. If the supply frequency is f_1 and number of poles of the motor is P, then the synchronous speed of rotating magnetic field is given by

$$N_s = \frac{120 f_1}{P} \tag{2}$$

The frequency of the induced emf in the rotor is given by

$$f_2 = \frac{\text{Poles} \times \text{Relative speed between the rotating flux and the rotor}}{120} \tag{3}$$

The stator and rotor must have same number of poles. In the wound rotor the number of poles is made equal by proper winding design. In squirrel cage rotor, the same number of poles is automatically produced by electromagnetic induction. Hence from equation 3,

$$f_2 = \frac{P(N_s - N_r)}{120} = \frac{P_s N_s}{120} = s\, f_1 = f_r \tag{4}$$

The synchronous speed of rotating magnetic field (clockwise), produced by rotor winding, *with respect to rotor* is given by

$$N_{s\,ROTOR} = \frac{120\, s\, f_1}{P} = sN_s \tag{5}$$

From equation 1, the speed of rotor structure (clockwise) *with respect to stator* is

$$N_r = (1 - s)N_s \tag{6}$$

Therefore, the actual speed of rotor rotating magnetic field (clockwise) *with respect to stator* is

$$sN_s + N_r = sN_s + (1 - s)N_s = N_s \tag{7}$$

Therefore, both the stator and rotor rotating magnetic fields rotate at the same speed N_s with respect to stator. So they are stationary with respect to each other. The mmfs of stator and rotor are stationary with respect to each other. Due to this a uniform torque is produced at all possible rotor speeds.

Further it can be concluded that any quantity of the rotor at slip frequency sf_1 when referred to the stator reacts at stator frequency f_1. Due to this it is possible to draw the load phasor diagram.

3.4 Analogy between Induction Motor and Transformer

An induction motor is similar to a transformer in many aspects. Both have primary and secondary windings. In induction motor, primary winding is a stator winding and secondary winding is a rotor winding. The mutual flux links both the windings. Due to this flux the emfs are induced in both the windings. Similar to transformer both the windings have leakage fluxes. Because of this, the induction motor is often called *generalized technology*. Since the rotor winding of an induction motor is short circuited the voltage and current equations of an induction motor are similar to those of a transformer with its secondary short circuited. Therefore, the voltage equation for a stator winding is

$$V_1 = E_1 + I_1(r_1 + jx_1) \tag{8}$$

The voltage equation for a rotor winding, at the time of start, is

$$V_2 = E_2 - I_2(r_2 + jx_2) = 0 \tag{9}$$

The voltage equation for a rotor winding, under running condition, is

$$sE_2 - I_2(r_2 + jsx_2) = 0 \tag{10}$$

Again similar to transformer, the equation for no load current is

$$I_0 = I_m + I_l \tag{11}$$

The no load current in induction motor varies from 30 to 50% whereas in transformer it varies from 2 to 6% of full load current. This is because of the fact that the mutual flux in the induction motor has to cross the air gap between stator and rotor. For this more mmf is required and ultimately more no load current.

3.5 Equivalent Circuit Diagram

The equivalent circuit of the induction motor can be derived exactly in the same way as that of transformer. The only difference is on account of the fact that the secondary winding, i.e., rotor winding of an induction motor rotates and therefore gives mechanical power as a output power. The rotor or secondary quantities can be referred to stator or primary side, similar to transformer. All the equivalent circuit parameters are expressed on per phase basis.

3.5(a) Stator Equivalent Circuit

On the basis of equations 8 and 11, the stator equivalent circuit diagram can be drawn as shown in Fig. 6.5. The stator terminal voltage V_1 has to overcome the counter emf E_1 and stator leakage impedance $r_1 + jx_1$. The stator current I_1 consists of two components, one no load current I_0, and another is load current component I_1' or rotor current component referred to stator side. The no load current I_0 consists of two components, one magnetizing current component I_m to produce the flux and second core loss component I_c. In case of transformer, the no load current gives only core loss but in case of induction motor, this current gives core loss and friction and windage loss.

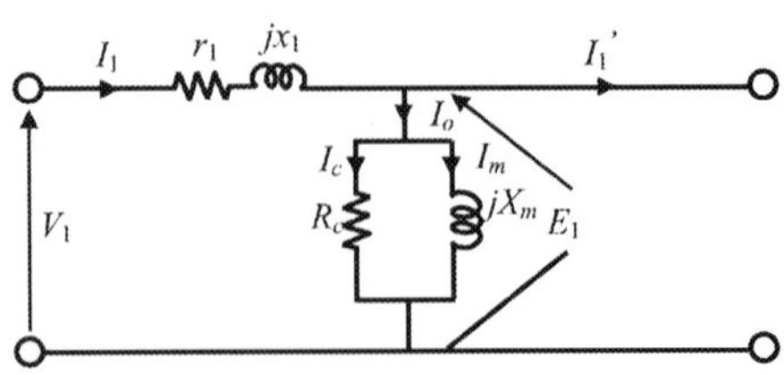

Fig. 6.5. Stator equivalent circuit diagram.

3.5(b) Rotor Equivalent Circuit

From equation 10, the rotor current can be written as

$$I_2 = \frac{sE_2}{r_2 + jsx_2}, \text{rotor current at slip frequency, } sf_1 = f_2 \tag{12}$$

$$= \frac{E_2}{\frac{r_2}{s} + jx_2}, \text{ rotor current at frequency, } f_1 \tag{13}$$

The magnitude and phase of I_2 obtained in equation 12 and 13 are same. But equation 12 gives the current of slip frequency f_2, due to sE_2 and equation 13 gives the current of frequency f_1, due to E_2. Depending upon these two equations, the equivalent circuit diagrams are drawn, as shown in Fig. 6.6.

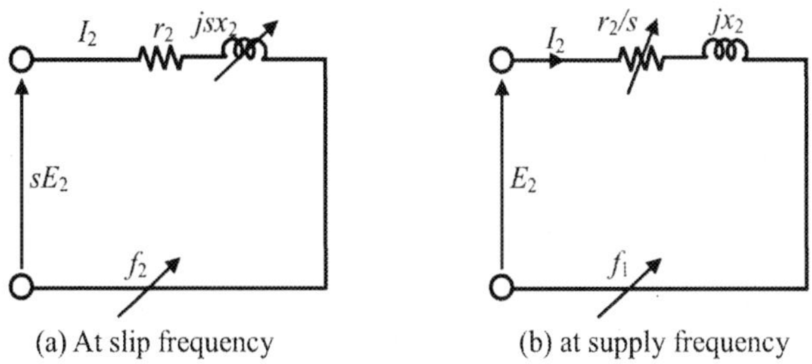

Fig. 6.6. Rotor equivalent circuit diagram.

3.5(c) Complete Equivalent Circuit

The rotor voltage E_2, the current I_2, resistance r_2/s and leakage reactance x_2 can be transferred to stator similar to transformer. So $E_1 = E_2/k$, $I_1' = kI_2$, $r_2'/s = (r_2/s)/k^2$ and $x_2'/s = x_2/k^2$. The resistance r_2'/s can be written as

$$r_2'/s = r_2' + r_2'\left(\frac{1-s}{s}\right) \tag{14}$$

The complete equivalent circuit diagram is shown in Fig. 6.7. In this figure all the rotor quantities are referred to stator. All the quantities are at stator frequency.

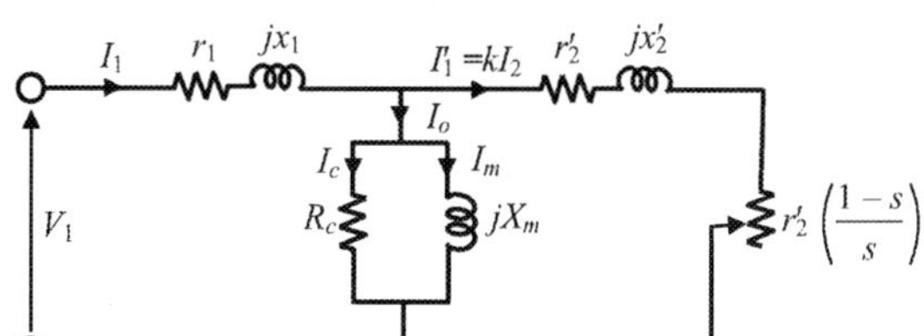

Fig. 6.7. Exact equivalent circuit diagram of a 3-phase induction motor referred to stator.

The equivalent circuit diagram shows the similarity between induction motor and transformer. At standstill $s = 1$, the circuit is equivalent to *short circuited* two-winding transformer. On no load, speed is near to synchronous speed, $s = 0$, the circuit is equivalent to *open circuited* transformer. The variable resistance $r_2'(1-s)/s$ is the electrical equivalent of the *variable mechanical load.*

3.6 Power and Torque

The power input to the stator is given by

$$P_1 = 3V_1 I_1 \cos\theta_1 \tag{15}$$

Some part of the power is consumed by stator winding (copper loss) and stator core (core loss). The remaining power is available at the air gap. Therefore the *total* air gap power is given by

$$P_g = P_1 - 3\, I_1^2 r_1 - P_c \tag{16}$$

The *per phase* air gap power can also be written from the rotor equivalent circuit diagram, as

$$P_g = E_2 I_2 \cos\theta_2 = E_2 I_2 \frac{r_2 / s}{\sqrt{(r_2 / s)^2 + x_2^2}} = I_2^2 \frac{r_2}{s} \tag{17}$$

Let us consider the referred parameters for writing the air gap power equation.

$$P_g = \frac{E_2}{k} I_2 k \frac{{r_2 / s}\Big/{k^2}}{\sqrt{(r_2 / s)^2 + x_2^2}\Big/{k^2}} = I_2^2 \frac{r_2}{s} = I_2^2 k \frac{2r_2}{k^2\, s} = I_1'^2 \frac{r_2'}{s} \tag{18}$$

$$= I_1'^2 r_2' + I_1'^2 r_2' \frac{(1-s)}{s} \tag{19}$$

= *Rotor copper loss + Internal mechanical power developed in the rotor*

$$= P_{rcu} + P_m$$

$$= sP_g + (1-s)P_g \tag{20}$$

If the iron loss in the rotor and friction and windage loss are subtracted from P_m, then the output power available at the shaft P_{sh} is obtained. On the name plate of motor, output power P_{sh} is given in kW or HP.

The *internal* or *gross torque developed* per phase, T_e is defined as the ratio of internal mechanical power developed in rotor, P_m to the rotor speed in mechanical radian per second, ω_{rMech}.

$$T_e = \frac{P_m}{\omega_{rMech}} \text{ watt/rad/sec (Nm)} = \frac{(1-s)P_g}{(1-s)\omega_{sMech}} = \frac{P_g}{\omega_{sMech}} = \frac{1}{\omega_{sMech}} I_1'^2 \frac{r_2'}{s} \quad (21)$$

Due to ω_{sMech}, the torque T_e can also be expressed in synchronous watts.

The output torque is $$T_{sh} = \frac{P_{sh}}{\omega_{rMech}} \quad (22)$$

3.7 Torque-Slip Characteristics

Under normal operating conditions of constant voltage and frequency, core loss is usually constant. In view of this, core loss resistance R_c can be omitted from the equivalent circuit, as shown in Fig. 6.8(a). In case of transformer, for simplification, the no load current branch is shifted towards voltage V_1 terminals because the no load current is 2 to 6% of full load current. But in case of induction motor, it will introduce some error since the no load current is 30 to 50% of full load current.

Simplification of this circuit can be obtained by *Thevenin's theorem* for easier computation. Consider two points *A* and *B*. Apply *Thevenin's theorem* viewing from the points A and B towards the voltage source V_I. Then the circuit consisting of r_1, x_1, X_m and source voltage V_1 can be replaced by an equivalent voltage source V_{Th} and an equivalent series impedance $Z_{Th} = R_{Th} + jX_{Th}$ as shown in Fig. 6.8(b).

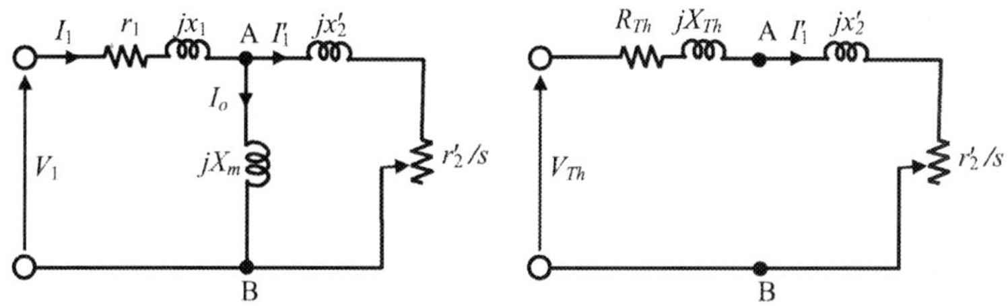

(a) Without core loss (b) Thevenin's equivalent

Fig. 6.8. Equivalent circuit diagram.

$$V_{Th} = \frac{V_1}{r_1 + j(x_1 + X_m)} jX_m \quad (23)$$

$$Z_{Th} = \frac{(r_1 + jx_1)(jX_m)}{r_1 + j(x_2 + X_m)} = R_{Th} + jX_{Th} \quad (24)$$

$$I_1' = \frac{V_{Th}}{\left(R_{Th} + \frac{r_2'}{s}\right) + j(x_2' + X_{Th})} \quad (25)$$

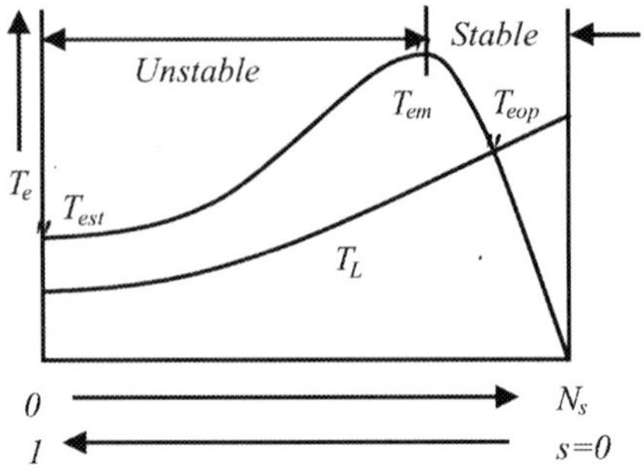

Fig. 6.9. Torque-slip characteristic.

From equation 21, the torque (per phase) can be written as

$$T_e = \frac{1}{\omega_{sMech}} I_1'^2 \frac{r_2'}{s} = \frac{1}{\omega_{sMech}} \cdot \frac{V_{Th}^2}{\left(R_{Th} + \frac{r_2'}{s}\right)^2 + \left(X_{Th} + x_2'\right)^2} \frac{r_2'}{s} \ Nm \tag{26}$$

Equation 26 is also called the equation for torque-slip characteristics.

The various points on the torque-slip (or speed) characteristics can be plotted from the above equation for different values of slip s and the general shape is shown in Fig. 6.9. The induction machine operates as motor when slip is positive. T_{est} is the starting torque, T_{em} is the maximum torque and T_{eop} is the operating point torque due to load torque T_L. From T_{est} to T_{em}, the operation of motor in unstable. Both torque and speed are increasing. From T_{em} to N_s, the operation of motor in stable. Increase in speed is restricted or opposed by decrease in torque. In this zone the characteristic is linear and is similar to DC shunt motor characteristic. Therefore this characteristic is, sometimes, called *shunt characteristic*. The slip is up to 5% in the stable zone.

3.7.1 Starting Torque T_{est}

At starting, $N_r = 0$ and slip $s = 1.00$. The starting torque T_{est} is given by

$$T_{est} = \frac{K_{Th}}{\left(R_{Th} + r_2'\right)^2 + \left(X_{Th} + x_2'\right)^2} \cdot r_2' \ Nm \tag{27}$$

Thus, the starting torque is *directly proportional* to the rotor circuit resistance r_2. According to maximum power transfer theorem ($Z_{load} = Z_{source}$), the maximum torque will occur when $\frac{r_2'}{s_{mt}} = \sqrt{R_{Th}^2 + X^2}$, where $X = X_{Th} + x_2'$, (see Fig. 6.8(b)). In order to get maximum starting torque, the rotor circuit resistance must be increased to $\sqrt{R_{Th}^2 + X^2}$. For slip = 1, $R_{add}' + r_2' = \sqrt{R_{Th}^2 + X^2}$. Therefore, the new value of rotor circuit resistance can be obtained by inserting external additional resistance, $R_{add}' = \sqrt{R_{Th}^2 + X^2} - r_2'$, in the rotor circuit at the time of starting.

3.7.2 Maximum Torque T_{em}

This is also called stalling torque, pull-out torque or breakdown torque. The slip at maximum torque is $s_{mt} = \dfrac{r_2'}{\sqrt{R_{Th}^2 + X^2}}$. The substitution of r_2'/s_{mt} in place of r_2'/s in equation 26 and simplification results in

$$T_{em} = \frac{K_{Th}}{2\left[R_{Th} + \sqrt{R_{Th}^2 + X^2}\right]} Nm \tag{28}$$

Thus the maximum torque is independent of rotor circuit resistance but the slip at maximum torque is directly proportional rotor circuit resistance.

3.7.3 Operating Torque T_{eop}

In the stable zone slip is small and r_2'/s becomes quite large as compared to other denominator terms of equation 26, so these terms can be ignored. Therefore,

$$T_{eop} = \frac{K_{Th}}{\left(\frac{r_2'}{s}\right)^2} \frac{r_2'}{s} = \frac{s\,K_{Th}}{r_2'} Nm \tag{29}$$

Thus in the stable zone the torque is directly proportional to slip and the characteristic is a linear one.

3.8 Behaviour under Load

The purpose of the load test is to study the behaviour of an induction motor under load. Sometimes this test is to perform the heat run test to confirm the rating of the motor. Suppose the motor is on no load. The slip is small, a no load current (30 to 50% of full load current) flows. As the mechanical load on the motor is increased, the speed drops due to the retarding effect and slip increases. So the rotor current and rotor power output increase. Thus the motor adjusts itself to the new load conditions of increased output, corresponding to which the input stator current also increases (stator flux decreases due to opposition of rotor flux, induced voltage in stator winding decreases) and speed drops a little. At light loads, the power factor is poor and lagging. Under load, it is lagging but improves, due to increase in active power (kW) and active component of the current while the reactive

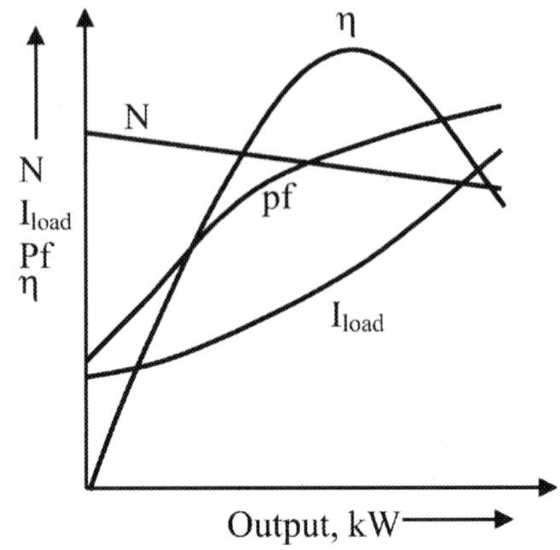

Fig. 6.10. Behaviour of an induction motor under load.

current component remains fairly constant. Under load, slip increases, the resistance equivalent to load $\{(1\text{-}s)/s\}r_2$, decreases and active component of current increases, due to negligible change in slip frequency, reactance remains fairly constant.

Thus, as the motor is loaded, the speed drops, current increases and power factor improves. Keeping the supply voltage constant, the variation of the different quantities like current, speed, power factor and efficiency with increase in load on the motor, are shown in Fig. 6.10.

4. APPARATUS REQUIRED

1. Three-phase induction motor: Rating (for example, 5 kW, 440 V, 1460 rpm).........
2. Coupled DC generator (shunt type): Rating (For example, 5 kW, 220 V)............
3. AC Ammeter: One, 0-20 amps.
4. AC Voltmeter: One, 0-300 or 600 volts.
5. DC Ammeter: One, 0-25 amps.
6. DC Voltmeter: One, 0-300 volts.
7. Wattmeter: One, 20A, 300 V.
8. Rheostat: One, 370 ohms, 2 amps.
9. Tachometer or speedometer: one.
10. Loading Rheostat: 5 kW, 230 V.

5. CIRCUIT DIAGRAM

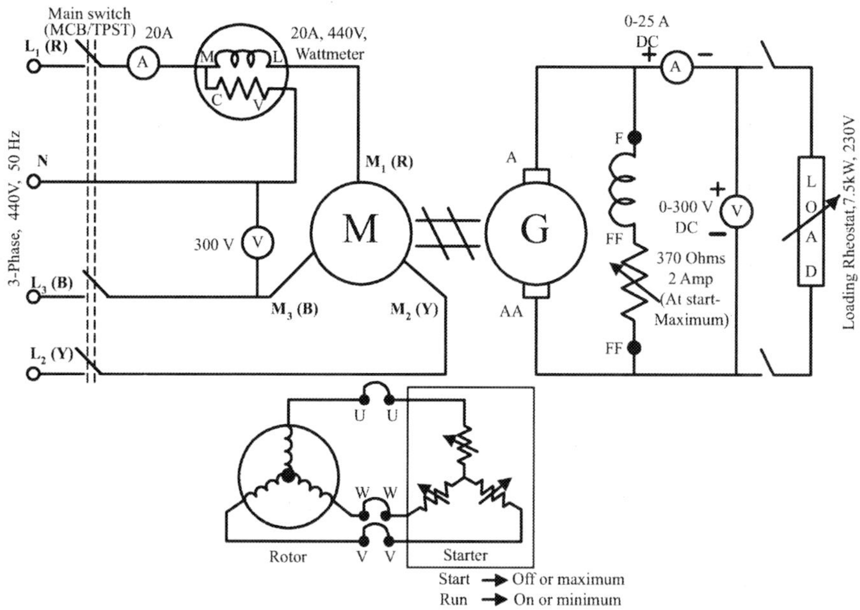

Fig. 6.11. Experimental set-up for load test on three-phase induction motor.

6. PROCEDURE

1. Note down the rating of the machines.
2. Make the connections as shown in Fig. 6.11.
3. Set up the generator field rheostat to maximum value (OFF).
4. Switch off all the loads. Put the starter to OFF position to have maximum resistance in rotor circuit, to produce more starting torque.
5. Switch on AC supply. Press green button of protective or current limiter box to start the motor.
6. Move the handle of starter from OFF to ON to decrease the resistance in rotor circuit and to run the motor to its normal speed (at minimum rotor resistance).
7. Excite the DC shunt generator to its rated voltage by decreasing field resistance (Off to On).
8. With no load, note down motor current, voltage, power, speed and generator current and voltage.
9. Put some load on the generator and bring generator voltage to its rated value and note down the above readings.
10. Take more readings by increasing the load gradually till the full load is reached.

7. OBSERVATIONS

Sr. No.	Observations						Calculations			
	Induction motor side				Generator Side		Gen output	Motor output	Motor efficiency	Ind. motor p.f.
	Voltage V_m	Current I_m	Power P_m	Speed r.p.m.	Voltage V_g	Current I_g	$V_g \times I_g$	$= \frac{V_g \times I_g}{0.85}$	$= \frac{Vg \times Ig}{0.85 \times 3 \times P_m}$	$= \frac{P_m}{V_m \times I_m}$
1.					Constant					
2.					↓					
..										
..										

8. RESULTS

The variation of load current, speed, power factor and efficiency with output of an induction motor has been plotted and studied. The characteristics are shown in Fig. 6.12.

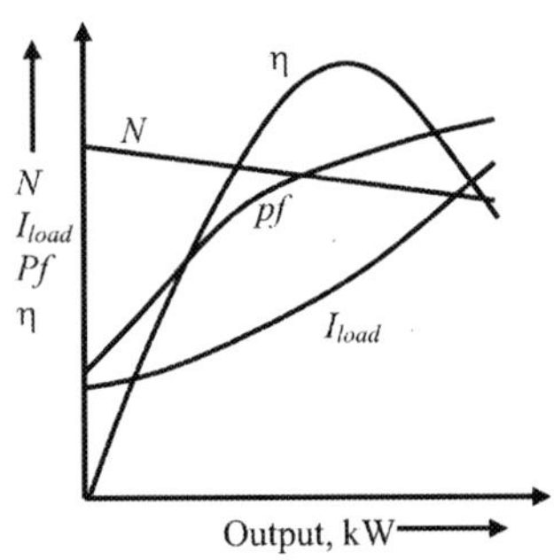

Fig. 6.12. Performance curves of a 3-phase induction motor under load.

9. CONCLUSION

1. The speed decreases as the load on motor increases.
2. The load current increases with an increase in output.
3. Power factor (lagging) improves with an increase in load.
4. The efficiency is zero at no load. It increases as the load increases and is maximum when variable losses are equal to constant losses.
5. The efficiency is maximum near full load.

10. POST-EXPERIMENTAL QUIZ

1. Why the power factor of an induction motor improves with increase in load?
2. Calculate the frequency of the supply depending on the no load speed?
3. Compare the SRIM and SCIM.

CHAPTER 7

To Determine Voltage Regulation and Efficiency of a Single-Phase Transformer by Direct Loading

1. OBJECTIVE

The objective of this experiment is to study the effects of load on the voltage and find the efficiency of the single-phase transformer.

2. PRE-EXPERIMENTAL QUIZ

1. What do you mean by voltage regulation?
2. Explain positive and negative voltage regulation.
3. Which parameter is responsible for voltage regulation?

3. THEORY

3.1 Voltage Regulation

Constant voltage is the requirement of most domestic, commercial and industrial loads. Therefore it is necessary that the output voltage of a transformer must stay within limit as the load and its power factor vary. After connecting the load, secondary current flows and there is voltage drop in the transformer due to the leakage reactances, winding resistances and load power factor. Due to this amount of voltage drop, secondary voltage decreases. The voltage regulation is used to identify the characteristic of the voltage change in a transformer with loading. Therefore, the transformers are designed in such a way to keep the leakage reactances and winding resistances as low as possible. The distribution transformer, which is directly connected to the consumer, is designed to have less than 5% voltage drop.

Voltage regulation of a transformer is defined as the arithmetical difference in the secondary terminal voltage between no load ($I_2 = 0$) and full rated load ($I_2 = I_{fl}$)

at a given power factor with the same value of primary voltage for both no load and rated load. It is expressed as either per unit or a percentage of the rated load voltage. The rated voltage is usually taken to be the nameplate value.

The *numerical difference* between no load and full load voltage is called ***inherent voltage regulation***.

$$\text{Inherent voltage regulation} = |V_{2nl}| - |V_{2fl}| \tag{1}$$

where $V_{2nl} = E_2$ = no load secondary voltage. As per IS, the secondary rated voltage is equal to the secondary voltage at no load.

V_{2fl} = rated secondary full load voltage with the same value of primary voltage for both no load and full load. (These are magnitudes, not phasors.)

Percentage voltage regulation at ***full*** load,

$$\%R = \left[\frac{|V_{2nl}| - |V_{2fl}|}{|V_{2fl}|}\right]_{|V_1|=\text{constant}} \times 100\ \% \tag{2}$$

Percentage voltage regulation at ***any*** load,

$$\%R = \left[\frac{|V_{2nl}| - |V_{2l}|}{|V_{2fl}|}\right]_{|V_1|=\text{constant}} \times 100\ \% \tag{3}$$

where V_{2l} = voltage at any load.

A distribution transformer should have good voltage regulation, i.e., small value of voltage regulation. A poor voltage regulation means more value of voltage regulation and this has a bad effect on the operation of tube lights, TV, motors, etc.

3.2 Calculation of Voltage Regulation Using Circuit Parameters

In the equivalent circuit diagram, R_2 is the total resistance of the transformer referred to secondary side and X_2 is the total leakage reactance of the transformer referred to secondary side. Therefore, $R_2 = r_2 + r_1k^2$ and $X_2 = x_2 + x_1k^2$.

Voltage regulation can also be calculated in terms of its circuit parameters. Since the voltage regulation is expressed in terms of secondary load voltage, the

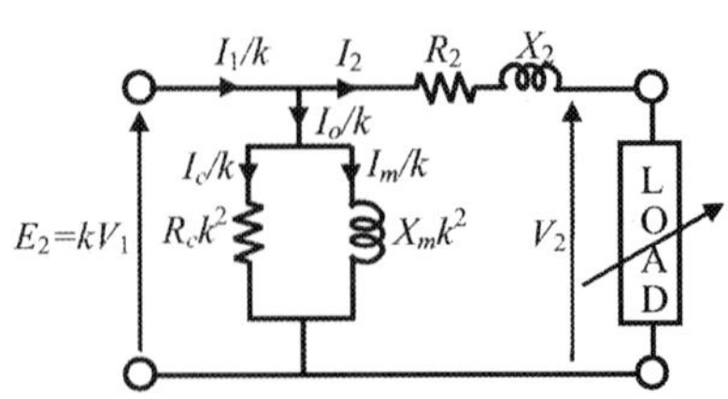

Fig. 7.1. Approximate equivalent circuit diagram of a transformer referred to secondary.

approximate equivalent circuit of the transformer referred to the ***secondary*** is considered for derivation of equations. The equivalent circuit of the transformer referred to the *secondary* is shown in Fig. 7.1. By KVL we can write

No load voltage, $E_2 = V_2 + I_2 Z_2$

Since $I_2 Z_2$ depends on the load power factor, the regulation depends on the load power factor. In order to calculate regulation the following steps are used:

1. Take V_2 as a reference phasor, $\therefore V_2 = V_2 \angle 0° = V_2 + j0$
2. Write I_2 in phasor form. For lagging power factor cos θ_2,

   $$I_2 = I_2 \angle -\theta_2 = I_2 \cos\theta_2 - jI_2 \sin\theta_2$$

   For leading power factor cos θ_2,

   $$I_2 = I_2 \angle +\theta_2 = I_2 \cos\theta_2 + jI_2 \sin\theta_2$$

   For unity power factor,

   $$I_2 = I_2 \angle 0 = I_2 + j0$$

3. Calculate total impedance, Z_2.
4. Calculate voltage drop and voltage regulation.

3.2(a) Voltage Regulation at Lagging Power Factor

The phasor diagram for the equivalent circuit diagram shown in Fig. 7.1, is drawn in Fig. 7.2. for a lagging power factor load. Here V_2 is taken as reference phasor. From the phasor diagram, we can write

$$\bar{E}_2 = \bar{V}_2 + \bar{I}_2 Z_2$$

$$= V_2 + j0 + (I_2 \cos\theta_2 - jI_2 \sin\theta_2)(R_2 + jX_2)$$

$$= (V_2 + I_2 R_2 \cos\theta_2 + I_2 X_2 \sin\theta_2) + j(I_2 X_2 \cos\theta_2 - I_2 R_2 \sin\theta_2) \qquad (4)$$

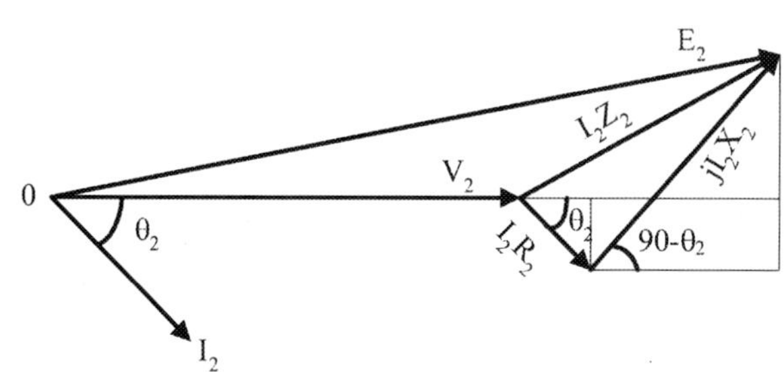

Fig. 7.2. Phasor diagram referred to secondary.

The phasor sum of V_2, I_2R_2 and I_2X_2 is E_2. Since the term $(I_2X_2 \cos \theta_2 - I_2R_2 \sin \theta_2)$ is small compared with the other term, it can be neglected. Therefore, the approximate value of E_2 is given by $E_2 = V_2 + I_2R_2 \cos\theta_2 + I_2X_2 \sin\theta_2$. The approximate voltage regulation is given by

$$\%R = \frac{E_2 - V_2}{V_2} \times 100\%$$

$$= \frac{(I_2R_2 \cos\theta_2 + I_2X_2 \sin\theta_2)}{V_2} \times 100\% \tag{5}$$

$$R \text{ in per unit} = \frac{I_2R_2 \cos\theta_2}{V_2} + \frac{I_2X_2 \sin\theta_2}{V_2} = \frac{R_2 \cos\theta_2}{V_2/I_2} + \frac{X_2 \sin\theta_2}{V_2/I_2}$$

$$R_{pu} = \frac{R_2}{Z_2}\cos\theta_2 + \frac{X_2}{Z_2}\sin\theta_2 = R_{2pu}\cos\theta_2 + X_{2pu}\sin\theta_2 \tag{6}$$

3.2(b) Voltage Regulation at Leading Power Factor

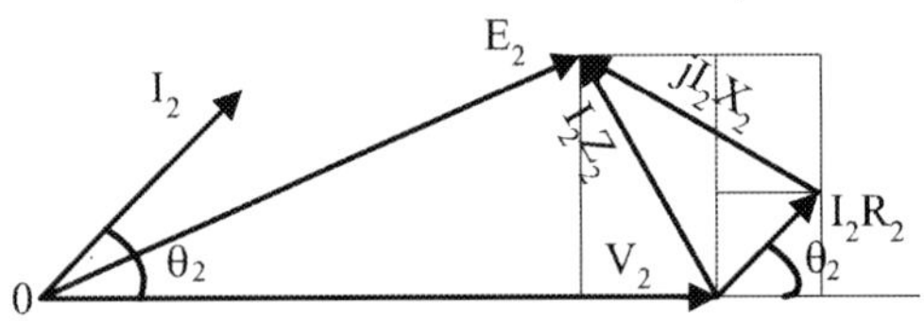

Fig. 7.3. Phasor diagram referred to secondary.

Fig. 7.3 shows the phasor diagram for leading power factor. Here again, V_2 is taken as reference phasor. The phasor sum of V_2, I_2R_2 and I_2X_2 is E_2. The approximate value of E_2 is given by $E_2 = V_2 + I_2R_2 \cos\theta_2 - I_2X_2 \sin\theta_2$. The approximate voltage regulation is given by

$$\%R = \frac{(I_2R_2 \cos\theta_2 - I_2X_2 \sin\theta_2)}{V_2} \times 100\% \tag{7}$$

$$R_{pu} = R_{2pu}\cos\theta_2 - X_{2pu}\sin\theta_2 \tag{8}$$

3.2(c) Voltage Regulation at Unity Power Factor

For unity power factor,

$$I_2 = I_2\angle 0 = I_2 + j0$$

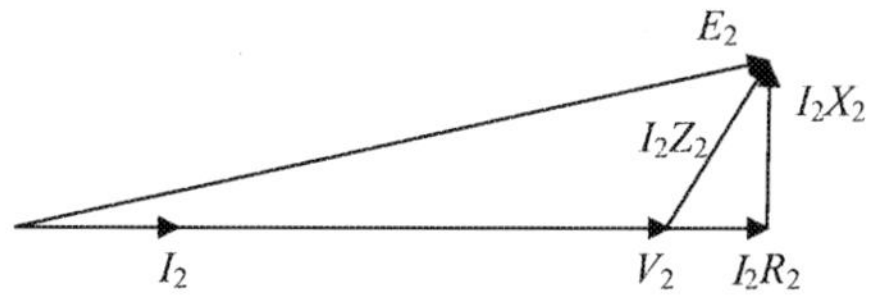

Fig. 7.4. Phasor diagram referred to secondary.

$$E_2 = V_2 + I_2Z_2 = V_2 + j0 + (I_2 + j0)(R_2 + jX_2) = (V_2 + I_2R_2) + jI_2X_2$$

The expression for E_2 at unity power factor can also be obtained from the expression for E_2 at lagging power factor by putting $\theta_2 = 0$. The voltage regulation is given by

$$\%R = \frac{\sqrt{(I_2R_2)^2 + (I_2X_2)^2}}{V_2} \times 100\% \approx \frac{I_2R_2}{V_2} \times 100\% \quad (9)$$

$$R_{pu} = R_{2pu} \quad (10)$$

3.2(d) Condition for Zero Voltage Regulation

The approximate voltage regulation is given by

$$\frac{E_2 - V_2}{V_2} = \frac{(I_2R_2 \cos\theta_2 + I_2X_2 \sin\theta_2)}{V_2}$$

For zero voltage regulation, $I_2R_2 \cos\theta_2 + I_2X_2 \sin\theta_2 = 0$

$$I_2R_2 \cos\theta_2 = -I_2X_2 \sin\theta_2,$$

$$\tan\theta_2 = -(R_2 / X_2) \ \therefore \theta_2 = -\tan^{-1}\left[\frac{R_2}{X_2}\right] \quad (11)$$

The negative sign indicates that *zero voltage regulation* occurs when the load is *capacitive*, that is, when the power factor is *leading*.

3.2(e) Condition for Maximum Voltage Regulation

For maximum voltage regulation, $\dfrac{d(Regulation)}{d\theta_2} = 0$

$$\frac{d}{d\theta_2} \frac{(I_2R_2 \cos\theta_2 + I_2X_2 \sin\theta_2)}{V_2} = 0$$

$$\therefore \tan\theta_2 = \frac{X_2}{R_2} \text{ and } \cos\theta_2 = \frac{R_2}{Z_2}$$

Thus, maximum voltage regulation occurs at lagging power factor of the load. The lagging power factor *angle of the load is equal to the angle* of the equivalent impedance of the transformer. The maximum value of voltage regulation is obtained by putting $\cos\theta_2$ and $\sin\theta_2$ in equation (6) and that is equal to Z_{2pu}.

$$R_{pu} = \frac{I_2 R_2}{V_2}\frac{R_2}{Z_2} + \frac{I_2 X_2}{V_2}\frac{X_2}{Z_2}$$

$$= \frac{I_2}{V_2 Z_2}(R_2^{\,2} + X_2^{\,2}) = \frac{I_2 Z_2}{V_2} = Z_{2pu} \tag{12}$$

Thus, the magnitude of maximum voltage regulation is equal to the *per unit value of equivalent leakage impedance of* the transformer. If the leakage impedance of the transformer is 5%, then the regulation of the transformer is 5%.

From the phasor diagrams it can be concluded that if $E_2 > V_2$ then voltage regulation is positive (lagging power factor), if $E_2 < V_2$ then voltage regulation is negative (leading power factor), and if $E_2 = V_2$ then voltage regulation is zero.

To determine voltage regulation, a given load is to be thrown off and two voltages (under load and after throwing the load off) are to be measured.

3.3 Efficiency

The ratio of the output power to the input power in a transformer is known as transformer efficiency η.

$$\eta = \frac{\text{Output power}}{\text{Input power}} = \frac{\text{Output power}}{\text{Output power + iron loss + copper loss}}$$

$$= \frac{V_2 I_2 \cos\theta_2}{V_2 I_2 \cos\theta_2 + P_c + I_2^{\,2} R_2} \times 100\%$$

The efficiency at full load is

$$\%\eta = \frac{V_2 I_{2fl} \cos\theta_2}{V_2 I_{2fl} \cos\theta_2 + P_c + I_{2fl}^2 R_2} \times 100\% \tag{13}$$

In transformer there are mainly two types of losses, namely: (i) core loss or iron loss, (ii) copper loss or ohmic loss. ***Iron loss*** occurs in the magnetic material of the transformer. This loss consists of: (i) hysteresis loss, and (ii) eddy current loss. These two losses are given by

$$P_h = k_h f B_{\max}^{1.6} v \text{ Watts and } P_e = k_e (f B_{\max})^2 v t^2 \text{ watts} \tag{14}$$

where k_h = Steimetz hysteresis coefficient, depends on the quality of the material

B_{max} = maximum flux density

v = volume of core, m^3

k_e = constant depending upon the electrical resistance of the magnetic material and grade of iron

t = thickness of the laminations of magnetic material.

Hysteresis loss is a measure of the electric energy required to overcome the retentivity of the iron in the magnetic flux path. Eddy current loss is due the flow of eddy current in the core. This loss varies as the square of the thickness of laminations. In order to reduce this loss thin laminations are used which are insulated from each other by a small layer of varnish.

The applied voltage V_1 can be written as

$$V_1 = E_1 = 4.44\phi_m f T_1$$

$$\therefore \qquad B_{\max} = \frac{V_1}{4.44 A_c f T_1} \tag{15}$$

$$\therefore \quad P_h = k_h f \left(\frac{V_1}{4.44 A_c f T_1} \right)^{1.6} v \text{ and } \therefore \; P_e = k_e \left(\frac{V_1}{4.44 A_c T_1} \right)^2 vt \tag{16}$$

The hysteresis loss is dependent on applied voltage and frequency. The eddy current loss is dependent on the square of the applied voltage and thickness of the lamination and independent of frequency.

Thus, core losses are dependent on flux density, applied voltage, frequency and thickness of the laminations. For a particular transformer V_1, f and t are constant. The flux density varies with the change in load but the variation is negligible. Therefore, the core losses are assumed to remain constant from no load to full load. These losses are determined from the ***open circuit test***.

The ***copper losses*** are the winding losses. The copper losses are present due to the resistance of the winding and current flowing through the winding, i.e., I^2r losses. The total copper losses in the transformer are the addition of primary and secondary copper losses. The copper losses are determined from the ***short circuit test.***

The ***stray loss*** and ***dielectric loss*** are usually small and therefore neglected. The stray load loss is due to the eddy current flow in the conductors, tanks. This eddy current is due to leakage flux. The dielectric loss occurs in insulating materials, that is, transformer oil and solid insulation of the transformer.

Since the transformer is a static device, there are ***no rotational losses*** such as windage and frictional losses in a rotating machine.

3.3.1 Condition for Maximum Efficiency

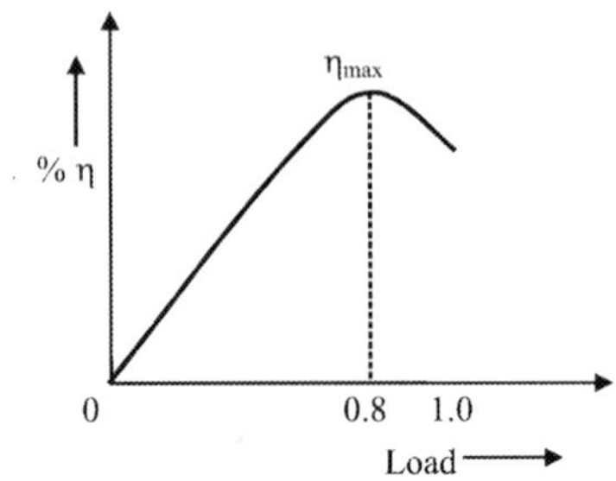

Fig. 7.5. Efficiency versus load curve.

For a constant applied voltage, flux remains constant and hence core loss remains constant. The copper loss varies as the square of load current or output. By the use of equation 13, the variation of efficiency with load is shown in Fig. 7.5. It is seen that the efficiency is low at small loads and reaches a maximum value for a certain load. The efficiency then decreases as the load is increased. The full load operating point is after the maximum efficiency point. Generally, the maximum efficiency occurs when the load on the transformer is from 80 to 90% of full load. In a well designed transformer the efficiency can be as high as 99%.

The efficiency is maximum when $d\eta/dt = 0$.

$$\therefore \frac{d\eta}{dt} = \frac{d}{dt}\left[\frac{V_2 I_{2fl}\cos\theta_2}{V_2 I_{2fl}\cos\theta_2 + P_c + I_{2fl}{}^2 R_2}\right] = 0 \tag{17}$$

$$\frac{\left[V_2 I_{2fl}\cos\theta_2 + P_c + I_{2fl}{}^2 R_2\right]V_2\cos\theta_2 - V_2 I_{2fl}\cos\theta_2\left[V_2\cos\theta_2 + 2I_{2fl}R_2\right]}{\left[V_2 I_{2fl}\cos\theta_2 + P_c + I_{2fl}{}^2 R_2\right]^2} = 0 \tag{18}$$

$$\left[V_2 I_{2fl}\cos\theta_2 + P_c + I_{2fl}{}^2 R_2\right]V_2\cos\theta_2 = V_2 I_{2fl}\cos\theta_2\left[V_2\cos\theta_2 + 2I_{2fl}R_2\right] \tag{19}$$

$$\therefore P_c = I_{2fl}{}^2 R_2 \tag{20}$$

Thus, the efficiency for a given power factor is maximum when the variable copper loss is equal to the constant iron loss.

3.3.2 Current and kVA at Maximum Efficiency

For maximum efficiency, variable copper loss = constant iron loss

$\therefore\ I_{2\max}^2 R_2 = P_c$ where $I_{2\max}$ is secondary current at maximum efficiency

$$I_{2\max}{}^2 = \frac{P_c}{R_2} = \frac{I_{2fl}^2 P_c}{I_{2fl}^2 R_2}$$

$$I_{2\max} = I_{2fl}\sqrt{\frac{P_c}{I_{2fl}^2 R_2}} \tag{21}$$

∴ Current at maximum efficiency

$$= \text{Full load current} \times \sqrt{\frac{\text{Constant iron loss}}{\text{Total F. L. copper loss}}}$$

Multiplying both sides by V_2, we get

$$\therefore\ V_2 I_{2\max} = V_2 I_{2fl}\sqrt{\frac{P_c}{I_{2fl}{}^2 R_2}}$$

$$\therefore\ VA_{\max eff} = VA_{full\,load}\sqrt{\frac{P_c}{I_{2fl}{}^2 R_2}}$$

$$\therefore\ \frac{VA_{\max eff}}{VA_{full\,load}} = \sqrt{\frac{P_c}{I_{2fl}{}^2 R_2}} = 80 \text{ to } 90\% \tag{22}$$

3.3.3 Efficiency Curves

The efficiency curve depends on load and its power factor. The maximum efficiency occurs at the same load current value I_{2max} regardless of the power factor of the load. The maximum efficiency for any power factor occurs at the same load and the highest possible efficiency occurs at unity power factor as shown in Fig. 7.6.

In indirect loading method, open circuit and short circuit tests are performed. The transformer parameters are calculated as per equivalent circuit diagrams and then the voltage regulation and efficiency can be calculated at different loads.

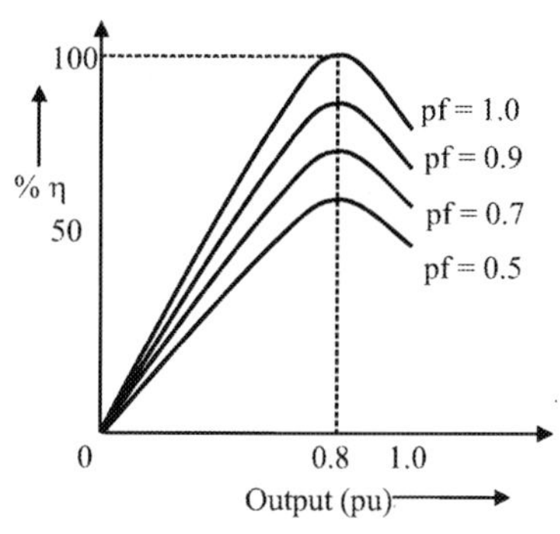

Fig. 7.6. Effect of power factor on efficiency.

In direct loading method, the transformer is loaded to its full load and then the voltage regulation and efficiency are calculated depending on actual operating conditions and hence actual results are obtained. It is obvious that more power is wasted in this test as compared to O.C. and S.C. tests as far as time is considered.

To determine the efficiency, power on the primary and on the secondary side has to be measured.

3.3.4 All Day (or Energy) Efficiency

The transformer efficiency discussed above is the ordinary, also called *commercial efficiency*, which is defined as the ratio of power output to power input.

There are certain types of transformers whose performance cannot be judged by commercial efficiency. Power transformers are connected in the system, when there is a load near to full load capacity of the transformer. During light load period, these transformers are disconnected. Therefore, the choice of a power transformer depends on its full load efficiency.

The distribution transformers are connected to the consumer terminals for 24 hours even when there is no load. Thus the core losses occur for the whole 24 hours and copper losses occur only when there is load. In view of this, the distribution transformers are designed to have very low value of core loss. The maximum efficiency may occur at about 50% of its rated output. Thus a distribution transformer should not be judged by its full load efficiency. The performance of a distribution transformer is more appropriately represented by *all day* or *energy efficiency* which is calculated on the basis of energy consumed during the whole day (24 hours).

All day efficiency is defined as the ratio of total energy output (kWh) for 24 hours to the total energy input for the same period.

$$\text{All day efficiency} = \frac{\text{Total energy output for 24 hours}}{\text{Total energy input for 24 hours}}$$

If the load variations are known for a particular time period, then the energy efficiency can be calculated for that time period of load cycle.

4. APPARATUS REQUIRED

1. Single phase transformer, rating: For example, 2.5 kVA, 230/400 V, with tappings.
2. Single phase variac suitable for transformer.
3. Ammeters: Two, 20 amps.
4. Voltmeters: Two, 300 volts.
5. Wattmeters: Two, 300 V, 20 A.
6. Load (Resistive): One, 5 kW, 220 V.
7. Load (Capacitive): One, 5 kVA, 415 V (line).

5. CIRCUIT DIAGRAM

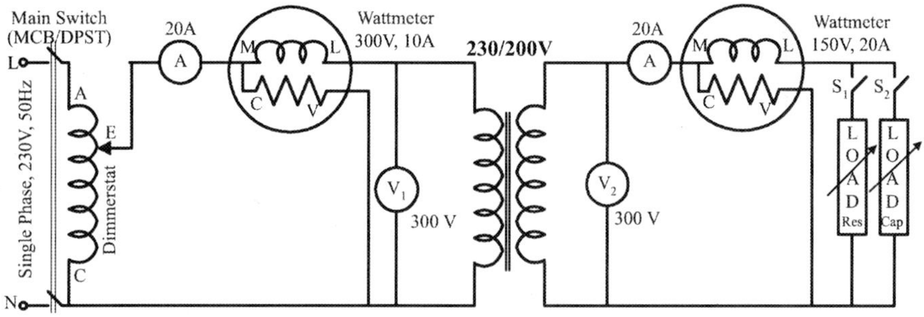

Fig. 7.7. Experimental set-up for direct loading on a single phase transformer

6. PROCEDURE

1. Make the circuit connections as shown in Fig. 7.7.
2. Put all the switches of loading rheostat in OFF position. Set the variac to zero output and switch on the supply.
3. Increase the voltage by variac, up to the rated voltage of the transformer.
4. Note down the meter readings under no load condition.
5. Increase the load by switching on one switch of loading rheostat and note down the meter readings.
6. Stepwise increase the load and note down the meter readings. Increase the load up to rated current of transformer.
7. Stepwise decrease the entire load. Switch off the supply. Disconnect the loading rheostat and connect the capacitor (secondary voltage should be equal to rated voltage of the capacitor): one reading for single capacitor and second reading for two capacitors in parallel.

7. OBSERVATIONS

	Observation						Calculations	
	Open circuit voltage E_2 = ... volts							
Sr. No.	Primary side			Secondary side			% Voltage regulation	% Efficiency
	Voltage V_p	Current I_p	Power P_p	Voltage V_s	Current I_s	Power P_s	$\%R = \frac{E_2 - V_s}{E_2} \times 100\%$	$\%\eta = \frac{P_s}{P_p} \times 100\%$
1.	Constant				0A			
2.	↓				1A			
..					..			
..					...			

8. RESULTS

The variation of output voltage V_s, voltage regulation % R, and efficiency % η with output of transformer has been plotted and studied. The characteristics are shown in Figs. 7.8 and 7.9 respectively.

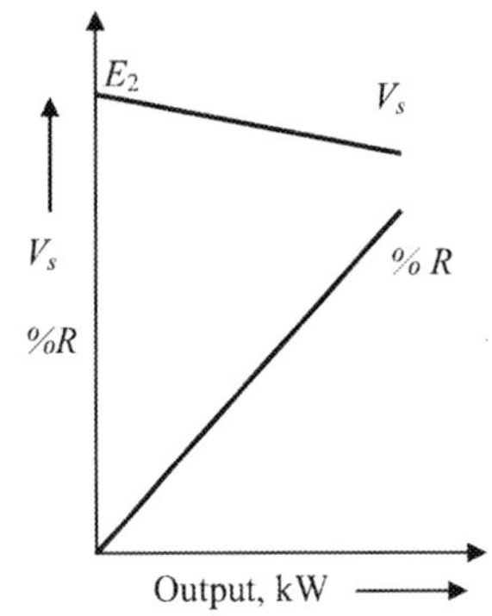

Fig. 7.8. Load voltage and % R versus output curve.

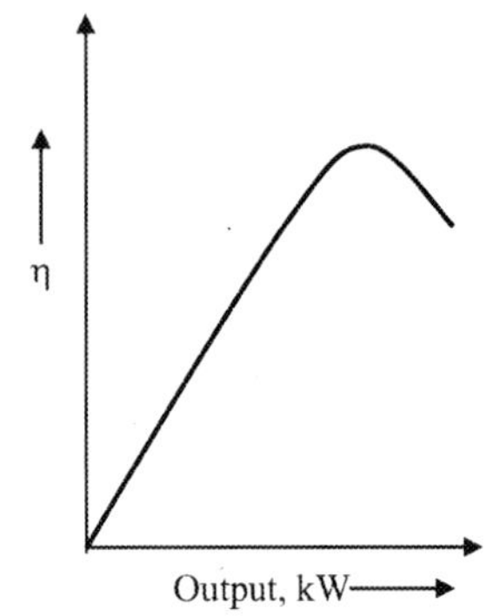

Fig. 7.9. Efficiency versus output curve.

9. CONCLUSION

1. The output voltage decreases as the output increases for rheostatic loading.
2. Regulation increases as the output increases for rheostatic loading.
3. For capacitive loading, output voltage is more than no load voltage.
4. The efficiency is zero at no load. It increases as the load increases and is maximum when variable losses are equal to constant losses.
5. The efficiency at rated output is less than the maximum value and the operating point is after the maximum efficiency point.
6. Maximum efficiency occurs when the load on the transformer is from 80 to 90% of full load.

10. POST-EXPERIMENTAL QUIZ

1. As the inductive load is increased, output voltage decreases, why ?
2. Draw the phasor diagrams of voltages and currents for inductive and capacitive load.
3. In the design of transformer, which precaution is taken to have better voltage regulation?

CHAPTER 8

To Study the Scott-Connection of Transformers (Three-Phase to Two-Phase Conversion)

1. OBJECTIVE

The main objectives of this experiment is to study

1. the conversion from three-phase to two-phase by using two single-phase transformers, and
2. the effect of balanced and unbalanced loading on three-phase system.

2. PRE-EXPERIMENTAL QUIZ

1. Explain Scott connection of transformers.
2. Name some specialized applications where two-phase supply is essential.
3. The neutral of three-phase system is connected to 28.8% tapping on teaser transformer. Why?

3. THEORY

Nowadays, three-phase supply is commonly used but for certain specialized applications two-phase supply is essential. Therefore, it is required to convert three-phase to two-phase. The applications of two-phase energy system are:

1. Single-phase arc furnaces. (Two single phase furnaces are used in together to draw balanced load from three-phase supply.)
2. Electric traction.
3. Low voltage single-phase rural supplies.
4. Two-phase control motors.
5. To link a three-phase system with a two-phase system with flow of power in either direction.

Charles F. Scott (1864–1944) converted two phases to three phases at Niagara Falls (Buffalo, NY, in 1896) and the arrangement used by him is known as Scott

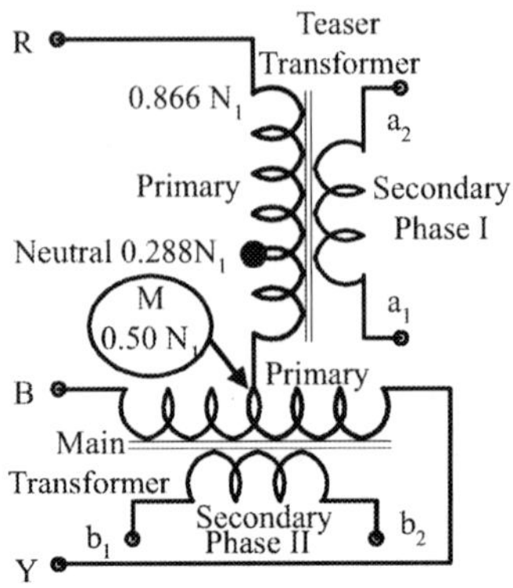

Fig. 8.1. Scott connection of transformers.

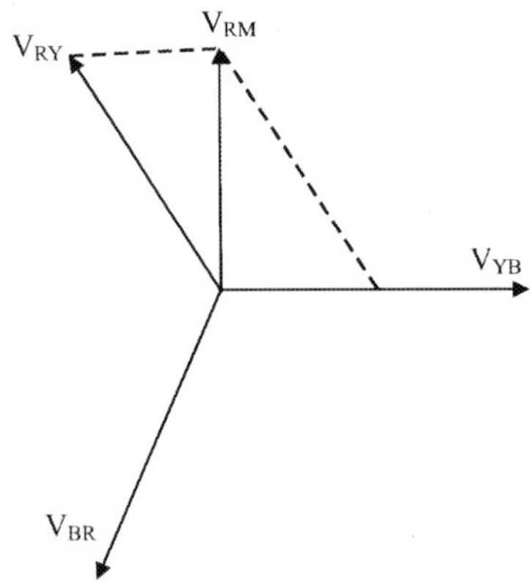

Fig. 8.2. Primary voltage phasor diagram.

connection, in his honour. So Scott connection is used to interlink two-phase and three-phase systems. Using this connection a three-phase supply can be converted to two-phase supply or vice versa. Two transformers are connected electrically but not magnetically. To achieve this, two single-phase transformers having **same number of secondary turns (N_2)** are used. One transformer is called **main transformer**, which is center tapped and the other is known as **auxiliary** or **teaser transformer** which has 0.866 N_1 turns. Primary winding of the main transformer is connected across two lines of three-phase supply (*Y* and *B*) and the primary of teaser transformer is connected across the third line (*R*) and the mid point *M* of the main transformer as shown in Fig. 8.1. The line voltages of the three-phase system V_{RY}, V_{YB}, V_{BR} which are balanced are shown in Fig. 8.2. The same voltages are shown as a closed equilateral triangle in Fig. 8.3(a).

$$|V_{RY}| = |V_{YB}| = |V_{BR}| = V$$

Let V_{YB} be taken as reference voltage so that

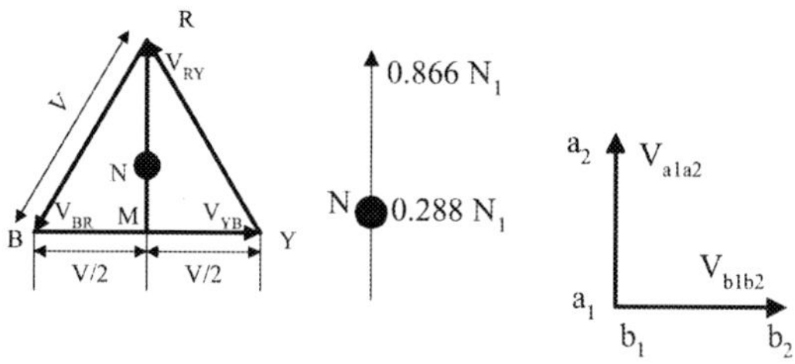

(a) Primary voltages (b) Secondary voltage

Fig. 8.3. Phasor diagrams for Scott connections.

$$V_{YB} = V\angle 0°,\ V_{RY} = V\angle +120°,\ V_{BR} = V\angle -120°$$

Since point *M* divides the primary *BY* of the main transformer in two equal halves, number of turns in portion *BM* is equal to the number of turns in portion *MY* which is equal to $(N_1)/2$. The voltages V_{BM} and V_{MY} are equal and they are in phase with V_{YB}.

$$\therefore\ V_{BM} = V_{MY} = \frac{1}{2}V_{YB} = \frac{1}{2}V\angle 0°$$

The voltage between *R* and *M* is $\bar{V}_{RM} = \bar{V}_{MY} + \bar{V}_{RY}$

$$\therefore\ V_{RM} = \frac{1}{2}V\angle 0° + V\angle 120° = \frac{1}{2}V + V\left(-\frac{1}{2} + j\frac{\sqrt{3}}{2}\right) = j\frac{\sqrt{3}}{2}V = 0.866V\angle 90°$$

Or by considering right angle triangle ΔRMY,

$$\therefore\ V_{RM} = \sqrt{V^2 - \left(\frac{V}{2}\right)^2} = \frac{\sqrt{3}}{2}V = 0.866V \qquad (1)$$

Thus, the voltage V_{RM} in the primary of teaser transformer is 0.866 of that in main transformer and leads it by 90° as shown in Fig. 8.2. In other words, the teaser transformer has a primary voltage rating, that is, 0.866 or 86.6% of the voltage rating of main transformer. (Even though input voltage of teaser is less than the input voltage of main, output voltages of both are same = teaser).This means voltage per turn in teaser transformer is V/N_1 which is the same as the voltage per turn of the main transformer. Hence, the voltage induced in the two secondary windings which have the same number of turns will be same in magnitude but they will be in phase quadrature.

The voltages belonging to a phase winding, on the same core, are drawn parallel to each other in the phasor diagram. Therefore, the secondary voltage V_{a1a2} is parallel to V_{MR} and V_{b1b2} is parallel to V_{YB}. Since V_{RM} leads V_{MB} by 90^0, V_{a1a2} (V_{2T}) leads V_{b1b2} (V_{2M}) by 90°.Thus a three-phase system can be converted to a two-phase system or vice versa.

Note that the role of two identical transformers can be interchanged. But they must have the tapping at 86.6% and 50%.

3.1 Position of Neutral Point N

The primary of the two transformers may have a four-wire connection to a three-phase supply if a tapping point *N* is provided on the primary of the teaser transformer such that

$$\text{Voltage across } RN = V_{RN} = \text{Phase voltage} = V/\sqrt{3} = 0.578V$$

Since the teaser voltage, equal to 0.866V is greater than 0.578V; the **neutral point** must be on the teaser primary. The neutral point is, therefore, *0*.866 V – 0.578 V= 0.288 V from *M*. The voltage V_{RM} = 0.866 V is divided into two parts; V_{RN} = 0.578 V and V_{MN} = 0.288 V such that the ratio

$$\frac{V_{RN}}{V_{MN}} = \frac{\text{Number of turns in } RN}{\text{Number of turns in } MN} = 2 \tag{2}$$

and the neutral point divides the teaser primary *RM* in the ratio *RN* : *MN* = 2:1.

3.2 MMF Balance

If exciting current is neglected, then the transformer action requires the balance between primary and secondary ampere-turns.

For main transformer, the transformation ratio *k*, is N_2/N_1.

Considering the secondary current $I_{b1b2} = I_{2M}$, primary mmf = secondary mmf

$$I_{YB} N_1 = I_{2M} N_2$$

$$I_{YB} = I_{2M} N_2/N_1$$

$$= k\, I_{2M} \tag{3}$$

The current from teaser transformer I_{RM} is divided into two equal currents I_{MY} and I_{MB}. The effects of these two currents are cancelled out.

For teaser transformer, the transformation ratio *k*, is $N_2/0.866\ N_1 = 1.15\ k$

Considering the secondary current $I_{a1a2} = I_{2T}$,

Primary mmf = secondary mmf

$$I_{RM}\,(0.866)\, N_1 = I_{2T} N_2$$

$$I_{RM} = I_{2T} N_2/0.866\ N_1$$

$$= 1.15\ k\ I_{2T} \tag{4}$$

3.3 Balanced Load

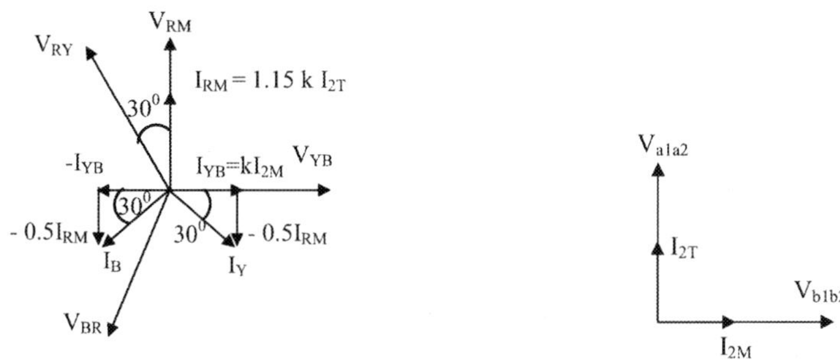

(a) Primary side (b) Secondary side

Fig. 8.4. Balanced load phasor diagram.

For simplicity consider $N_2/N_1 = 1$. For unity power factor balanced load, I_{2M} and I_{2T} are drawn on secondary side. On primary side, as per equation 4, I_{RM} is drawn in parallel to I_{2M}. I_R is the R-phase current entering into the 86.6 tapping point.

$$\therefore \; \bar{I}_R = 1.15 I_{2T} \tag{5}$$

Current I_Y is the phasor sum of I_{YB} and $-0.5I_{RM}$.

$$\therefore \; \bar{I}_Y = \bar{I}_{YB} - 0.5\bar{I}_{RM} = \bar{I}_{YB} - 0.5(1.15)I_{2T}$$

$$= \bar{I}_{2M} - 0.5(1.15)j\bar{I}_{2M} = \bar{I}_{2M}(1 - j0.575)$$

$$= 1.15\,\bar{I}_{2M}\angle -30° \tag{6}$$

Similarly, for current I_B

$$\therefore \; \bar{I}_B = -\bar{I}_{YB} - 0.5\bar{I}_{RM} = -\bar{I}_{YB} - 0.5(1.15)I_{2T}$$

$$= -\bar{I}_{2M} - 0.5(1.15)j\bar{I}_{2M} = \bar{I}_{2M}(-1 - j0.575)$$

$$= 1.15\bar{I}_{2M}\angle +30° \tag{7}$$

Thus, for a balanced two-phase load, the primary currents are also balanced. This is true whether the load power factor is unity or not.

3.4 Unbalanced Load

Suppose secondary load is unbalanced. The secondary currents I_{2T} and I_{2M} are not equal in magnitudes and are lagging by θ_1 and θ_2 to their respective voltages.

On primary side, as per equation 4, I_{RM} is drawn in parallel to I_{2M}. I_R is the R-phase current entering into the 86.6 tapping point.

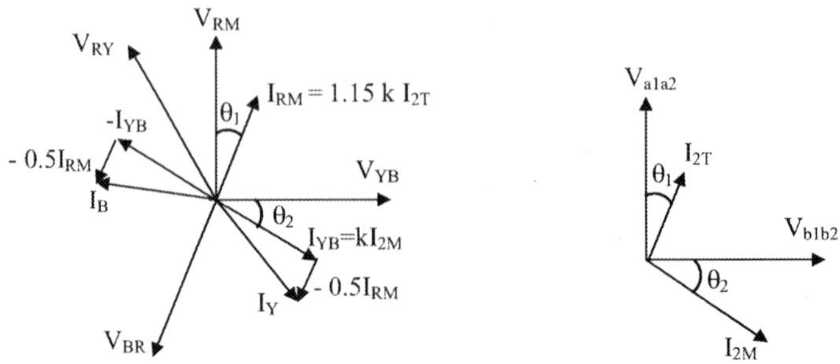

(a) Primary side (b) Secondary side

Fig. 8.5. Unbalanced load phasor diagram.

$$\therefore \ \bar{I}_R = 1.15 I_{2T} \tag{8}$$

This current makes an angle $30 + \theta_1$ with line voltage V_{RY}. The current I_Y is the phasor sum of I_{YB} and $-\,0.5\ I_{RM}$.

$$\therefore \ \bar{I}_Y = \bar{I}_{YB} - 0.5\bar{I}_{RM} = \bar{I}_{YB} - 0.5(1.15)I_{2T}$$

$$= \bar{I}_{2M} - 0.5(1.15)\bar{I}_{2T} \tag{9}$$

Similarly, for current I_B

$$\therefore \ \bar{I}_B = -\bar{I}_{YB} - 0.5\bar{I}_{RM} = -\bar{I}_{YB} - 0.5(1.15)I_{2T}$$

$$= -\bar{I}_{2M} - 0.5(1.15)\bar{I}_{2T} \tag{10}$$

Thus, for a unbalanced two-phase load, the primary currents are also unbalanced.

3.5 Advantage

The single phase loading or unbalanced loading on Scott connection results in less unbalance than the single phase loading on three-phase system.

4. APPARATUS REQUIRED

1. Single-phase transformers: Two (one main and one teaser), rating: for example 2.5 kVA.
2. Three-phase variac suitable for transformers.
3. Ammeters: Five, 10 amps or multimeter or clamp meter.

4. Voltmeters: Two, 300 volts and one of 600 V or multimeter or clamp meter
5. Loading rheostats: Two, 5kW, 220V.

5. CIRCUIT DIAGRAM

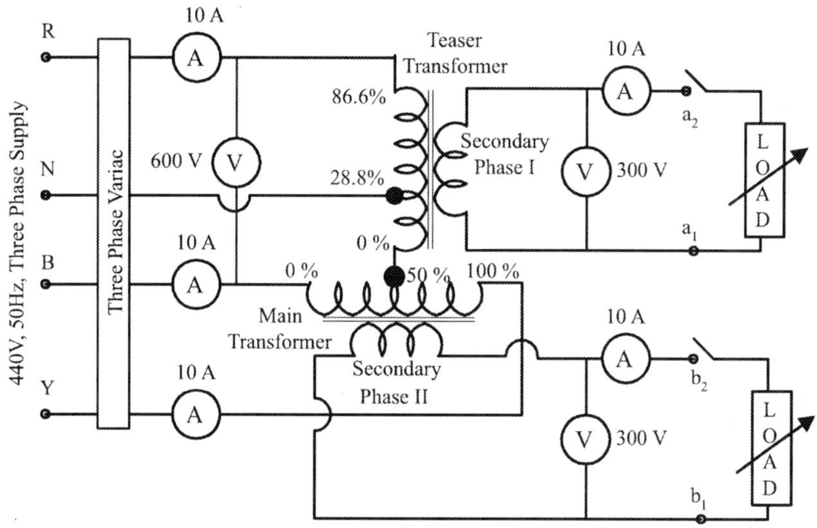

Fig. 8.6. Experimental set-up for the study of Scott connection of transformers.

6. PROCEDURE

A. Balanced Operation

1. Make the connections as shown in Fig. 8.6.
2. Put all the switches of loading rheostat in OFF position. Set the variac to zero output and switch on the supply.
3. Increase the voltage by variac, up to the rated voltage of the transformer.
4. Note down the meter readings under no load condition.
5. Symmetrically switch on the load on the secondary sides and note all the readings.
6. Repeat step 5 up to rated current.

B. Unbalanced Operation

1. Switch OFF all the loads.
2. Switch ON the load on the secondary of the main transformer.
3. Note down the meter readings under this condition.
4. Increase the load and note down the meter readings.
5. Switch OFF the load on the secondary of main transformer and switch ON the load on the secondary of teaser transformer.
6. Increase the load and note down the meter readings.

7. OBSERVATIONS

A. Balanced Operation

Sr. No.	Primary line currents			Primary line voltage	Secondary currents		Secondary voltages	
	I_1 A	I_2 A	I_3 A	V_L	I_{MAIN} A	I_{TEASER} A	V_{MAIN} V	V_{TEASER} V
1.					2 A	2 A		
2.					4 A	4 A		
.					.	.		
..					..	..		

B. Unbalanced Operation

Sr. No.	Primary line currents			Primary Line Voltage	Secondary currents		Secondary voltages	
	I_1 A	I_2 A	I_3 A	V_L	I_{MAIN} A	I_{TEASER} A	V_{MAIN} V	V_{TEASER} V
1.					0 A	2 A		
2.					0 A	4 A		
.					2 A	0 A		
..					4 A	0 A		

8. CALCULATION

Under balanced condition, the phasor sum, $\bar{V}_{MAIN} + \bar{V}_{TEASER} = \sqrt{2}\,V_{MAIN} = \sqrt{2}\,V_{TEASER}$.

9. RESULTS

The differences between three-line currents on primary side under balanced and unbalanced conditions are studied.

10. CONCLUSION

1. Three-phase system can be converted to two-phase system by using Scott connections.

2. The secondary voltages are in phase quadrature, then under balanced condition, $V_{MAIN} = V_{TEASER}$ and the phasor sum is $\overline{V}_{MAIN} + \overline{V}_{TEASER} = \sqrt{2}V_{MAIN} = \sqrt{2}V_{TEASER}$.
3. Single phase loading or unbalanced loading on Scott connection results in less unbalance than the single-phase loading on three-phase system.

11. POST-EXPERIMENTAL QUIZ

1. Draw the waveforms of output voltages.
2. Show that $\overline{V}_{MAIN} + \overline{V}_{TEASER} = \sqrt{2}V_{MAIN} = \sqrt{2}V_{TEASER}$.
3. What do you mean by teaser?

CHAPTER 9

To Study the Variation of Speed and Load Test on Schrage Motor

1. OBJECTIVE

The main objectives of this experiment are:

1. to see the effects of brush angle on the speed of the Schrage motor.
2. to study the effects of load on the efficiency, power factor and the output quantity, i.e., speed of the Schrage motor.

2. PRE-EXPERIMENTAL QUIZ

1. How many windings are there in Schrage motor and where they are housed?
2. How much is the frequency of the voltage induced in the tertiary winding?
3. Compare induction motor and Schrage motor.

3. THEORY

It has been seen that an induction motor has substantially constant speed in its operating range. Even though it is cheap, rugged in construction and requiring minimum maintenance, it has low starting torque. The starting torque of slip ring induction motor can be increased by increasing rotor circuit resistance. But otherwise, AC commutator motors are the other option for induction motors. The problem of AC commutator motors is of its medium ratings due to commutation problem. Now, with the inventions in power electronics, the frequency converting action has been taken over by it. Though AC commutator machines can still be found in older installation, they have for all practical purposes become a part of the history of electrical machines. The AC commutator machine can be operated near to unity power factor and at wide range of speed control. Thus, economy can be achieved. To understand the principle of operation and basic constructional features, this experiment on Schrage motor is performed.

3.1 Commutator-Frequency Converter

To understand the working principle, consider the schematic diagram of a two-pole DC machine, as shown in Fig. 9.1(a). The rotation of armature in the field flux gives rise to rotational emf E across the q-axis ***brushes***, whose magnitude is given by

$$E = \frac{\Phi ZN}{60}\frac{P}{a} \text{ Volts} \tag{1}$$

Now rotate the ***field poles*** clockwise through 90°, to the position shown in Fig. 9.1 (b). At this position emf E is reduced to zero. A further rotation through 90°, as shown in Fig. 9.1 (c), reverses the polarity of E but without changing its magnitude. If the poles are again rotated by 90° clockwise, as shown in Fig. 9.1 (d), the emf is again reduced to zero. The variation of E due to the rotation of field poles is shown in Fig. 9.1(e).

If the speed of field poles is n rps, then the frequency of the emf E at the brushes is given by

$$f_b = \frac{Pn}{2} = n \text{ Hz, for 2 poles} \tag{2}$$

If the brushes are rotated in the same speed and same direction of poles, then the relative speed between poles (or field) and brushes will be zero and the frequency of the brush emf will be zero (DC).

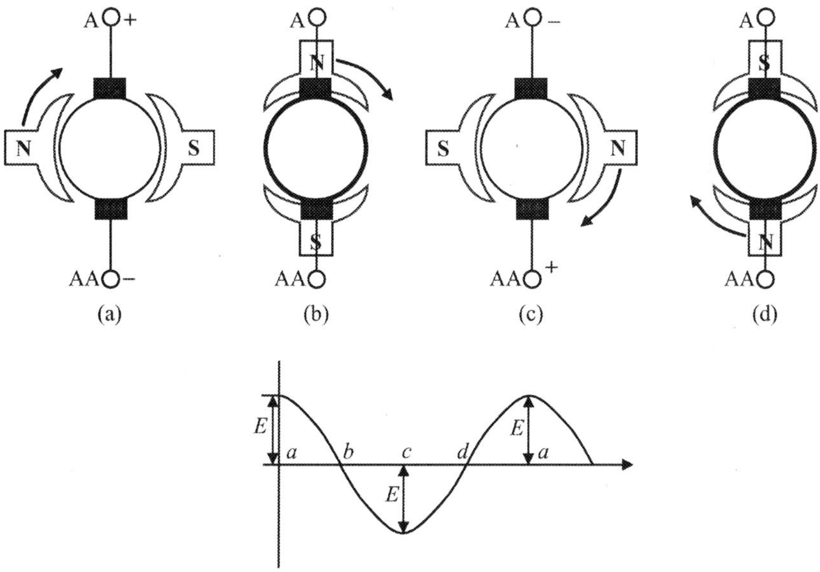

(e)

Fig. 9.1. Rotating frequency converter.

Now if the speed of armature field is n_r (with brushes stationary), the speed of poles is n and both are rotating in opposite direction, then the relative speed between air gap field (due to armature and pole field) and brushes will be $n + n_r$ and the frequency of the brush emf will be

$$f_b = \frac{P(n+n_r)}{2} Hz \tag{3}$$

If the poles are rotated in the same direction of rotation of armature field, then the relative speed between air gap field and brushes will be $n - n_r$ and the frequency of the brush emf will be

$$f_b = \frac{P(n-n_r)}{2} Hz \tag{4}$$

The expression for emf is given by

$$E = K\frac{P(n \pm n_r)}{2} \tag{5}$$

The angle between brushes is $\theta = 180°$.

$$\therefore E = K\frac{P(n \pm n_r)}{2}\sin\frac{\theta}{2} \tag{6}$$

Thus, the magnitude of the brush emf can be increased by increasing the angle between the two brushes. The axis of brush emf can be shifted by shifting both the brushes simultaneously in the same direction. A set of three-phase emfs of adjustable magnitude and axis with movable brushes mounted on a commutator is shown in Fig. 9.2.

Consider inverted induction motor with primary winding on rotor. The three-phase supply is given to the rotor with the use of three slip rings. Due to this the

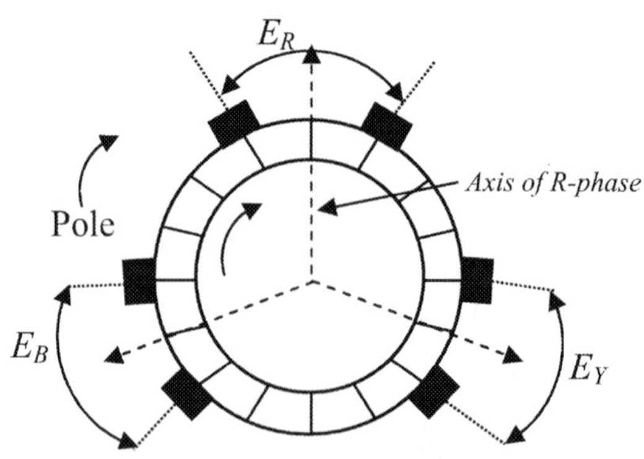

Fig. 9.2. Generalized frequency converter.

rotating magnetic field is set up, which rotates at synchronous speed n_s with respect to rotor. As per induction motor action, rotor starts rotating at a speed n_r in opposite direction to that of rotating magnetic field. If the rotor winding is connected to the commutator over which there are brushes, then the field speed in space or field speed relative to the brushes will be $n - n_r = sn_s$.

Therefore, brush frequency

$$f_b = \frac{P(n - n_r)}{2} = \frac{P s n_s}{2} = sf\ Hz \tag{7}$$

= slip frequency.

Thus, whatever may the speed of induction motor, the brush voltage frequency is always equal to the rotor frequency *sf* and is therefore suitable for injection into the secondary winding (stator in this case) of the motor.

3.2 Effect of Slip-Frequency Voltage

For induction motor, we have seen that the power transferred across the air gap from stator to rotor or *air gap power* is

$$P_g = I_1'^2 r_2' + I_1'^2 r_2' \frac{(1-s)}{s}$$

= Rotor copper loss + Internal mechanical power developed in the rotor

$= P_{rcu} + P_m$

If air gap power is multiplied by slip s then sP_g is the slip power. This slip power is positive when slip is positive, i.e., induction motor runs at sub-synchronous speed. But if slip power is made negative then the slip is negative and motor runs at super-synchronous speed. This slip power flow can be controlled by slip frequency voltage. The power factor can be controlled by shifting the axis of this slip frequency voltage.

Thus the speed and power factor of slip ring induction motor can be controlled by injecting *slip frequency voltage* in the rotor circuit.

3.3 AC Commutator Type Inverted Induction Motor or Schrage Motor

The method explained above of obtaining slip frequency emf and injecting it into the secondary winding of an induction motor is incorporated in a single machine called the Schrage motor. In 1911, K.H. Schrage of Sweden combined elegantly a SRIM (WRIM) and a frequency converter into a single unit.

Schrage motor is basically an inverted polyphase induction motor, with primary winding on the rotor and secondary winding on the stator. The primary winding on the rotor is fed through three slip rings and brushes at line frequency; secondary winding on the stator has slip frequency voltages induced in it. In addition to these two windings, there is a third winding called ***tertiary*** or ***adjusting*** or ***regulating winding***, which is housed in the same rotor slots of the primary, as shown in Fig. 9.4. The tertiary winding is connected to the commutator on which there are three sets of *movable* brush pairs, to collect the required emf for injection into the secondary circuit for speed and power factor control. The tertiary winding, near to air gap, have less leakage reactance, which helps for doing good commutation.

The secondary phase winding on stator are not connected to each other but are connected to brushes a, b, c, d, e, f as shown in Fig. 9.3. Alternate brushes a, c, e, 120 electrical degrees apart, are mounted on one brush rocker and brushes b, d, f on the second brush rocker. The angle between brushes ab, cd, and ef can be controlled by means of rack and pinion arrangement and one hand wheel provided outside the motor frame. Thus Schrage motor incorporates the function of polyphase induction motor and frequency converter.

If the primary and secondary windings are interchanged, the performance of three-phase induction motor is unaffected, but in a Schrage motor, it is essential to have primary winding on the rotor and secondary winding on the stator.

At standstill, if three-phase supply is given to primary winding, a magnetic field is produced, rotating at synchronous speed N_s with respect to primary and secondary conductors. Now as per induction motor action, rotor must rotate at a

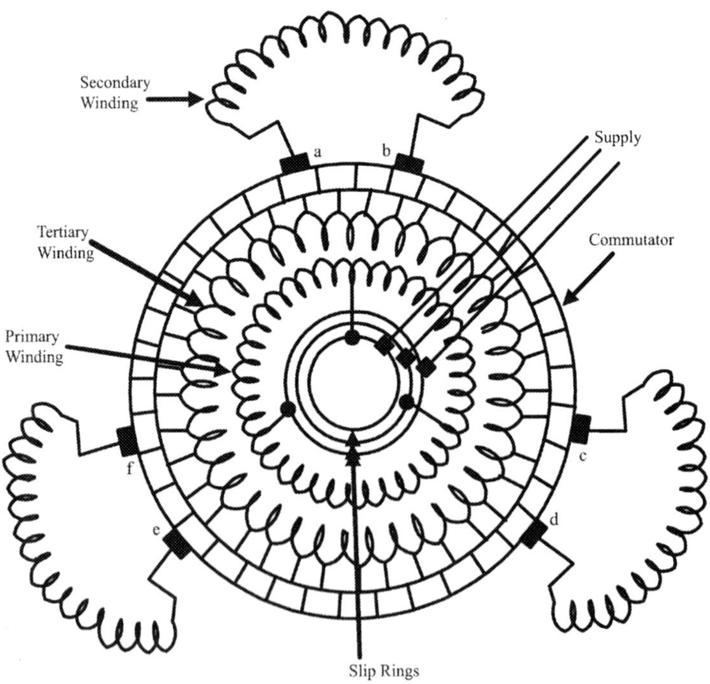

Fig. 9.3. Schrage motor: Schematic diagram.

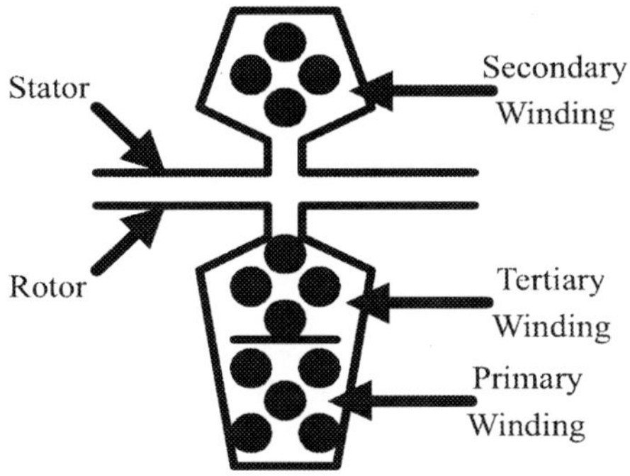

Fig. 9.4. Arrangement of three windings.

speed N_r, in a direction opposite to the synchronously rotating magnetic field since primary winding is on the ***rotor***. This opposes the flux cutting by the stator conductors. As a result of it, the rotating field speed in the air gap becomes $(N_s - N_r)$, i.e., slip speed. This air gap flux rotating at slip speed with respect to stator induces slip frequency emfs E_2 in the stator winding. Since the brushes are stationary, slip frequency emfs E_j are picked up by these brushes, as per equation 7. Due to the equal slip frequency, brush emfs are suitable for voltage injection into the secondary winding, i.e., stator winding.

The tertiary and primary windings, being in the same slots, are mutually coupled. Therefore, the emfs induced in the tertiary winding are by transformer action and are always at line frequency f, at all the rotor speed.

At the time of start, the brush pairs ab, cd, and ef are **shorted**, similar to **shorted** secondary winding of induction motor and the Schrage motor operates as an ordinary induction motor at sub-synchronous speed almost equal to synchronous speed, as shown in Fig. 9.5(a). The **shorted** secondary winding (brush angle zero) is also a condition for starting the Schrage motor. Now brushes

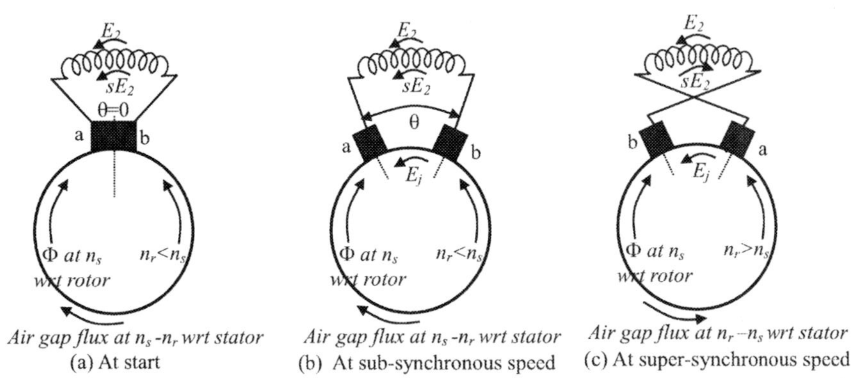

Fig. 9.5. Speed control.

a & b, c & d, e & f are separated by an angle θ such that the injected voltage E_j is opposite to E_2 and resultant secondary voltage decreases. Due to less secondary voltage current decreases, torque decreases and speed decreases and motor runs at sub-synchronous speed, as shown in Fig. 9.5(b).

When the motor runs at synchronous speed, magnetic field is stationary in space and tertiary winding acts as a DC armature and DC current flows in secondary winding. Thus, Schrage motor operates as a synchronous motor.

If the injected voltage E_j is in phase with E_2, resultant secondary voltage increases. Due to more secondary voltage current increases, torque increases and speed increases and motor runs at super-synchronous speed, as shown in Fig. 9.5(c).

Now let, brushes a & b, c & d, e & f be moved bodily, such that the two centers of E_2 and E_j are not coincident and have an angle ρ. This angle ρ, shifts the secondary current and reflected primary current in the phasor diagram, to control the power factor as shown in Fig. 9.6. Thus, speed and power factor are controlled by changing the angle between the brushes.

Components $E_j \cos \rho$ and $E_j \sin \rho$ are assumed to effect the speed and power factor respectively. The magnitude of E_j depends upon the brush separation angle θ, then the expression for E_j can be written as

$$E_j = E_{j\max} \sin\frac{\theta}{2} \tag{8}$$

The voltages E_2 and E_j (for fixed θ) remain constant but sE_2 changes as the rotor rotates. The voltage equation for the secondary circuit with E_j is given by

$$sE_2 \pm E_{j\max} \sin\frac{\theta}{2} = I_2 z_2 \tag{9}$$

Under no load condition, I_2 is small and $I_2 z_2$ can be neglected.

$$s_0 E_2 \pm E_{j\max} \sin\frac{\theta}{2} = 0$$

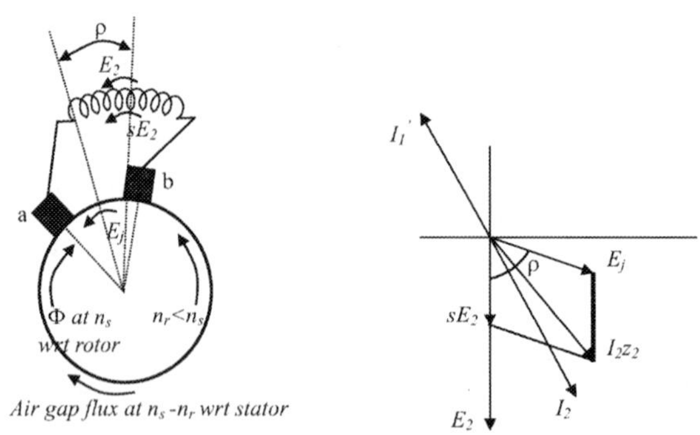

Fig. 9.6. Speed and power factor control.

$$\therefore s_0 = \mp \frac{E_{j\max}}{E_2} \sin\frac{\theta}{2} = \mp\, k \sin\frac{\theta}{2} \tag{10}$$

Thus, no load speed is

$$n_{r0} = n_s(1 - s_0) = n_s\left(1 \pm k\sin\frac{\theta}{2}\right) \tag{11}$$

where k is a constant whose value depends upon the turns ratio of the secondary and tertiary windings and θ is brush separation angle in electrical degrees. Plus sign indicates the speed above n_s and minus sign indicates the speed below n_s.

The lowest possible no load speed is $n_{r0.min} = n_s(1 - k)$ and is possible when $\theta = 180°$ electrical, i.e., when the brushes are *one pole pitch* apart. If the brush positions are now interchanged so that $\theta = -180°$ electrical, then maximum no load speed is $n_{r0.max} = n_s(1 + k)$. Thus, depending on the value of k, the speed can be varied over the range of $0.4n_s$ to $1.4n_s$. The no load speed with brush separation curve for a 4 pole, 50 Hz motor is shown in Fig. 9.7(a). The efficiency of the motor on load is slightly lower than that of a corresponding induction motor as this motor has additional components in the form of commutator and brushes and there are additional losses in the tertiary winding. This motor has increased efficiency at higher speed than at lower speed. The power factor also improves with the rise in speed. The typical efficiency and power factor curves are shown in Fig. 9.7(b) and (c) respectively.

3.3.1 Starting of Schrage Motor

At the time of starting, $s = 1$ and if $I_{2.st}$ is the starting current, then equation 9 becomes

$$E_2 \pm E_{j\max} \sin\frac{\theta}{2} = I_{2.st} z_2 \tag{12}$$

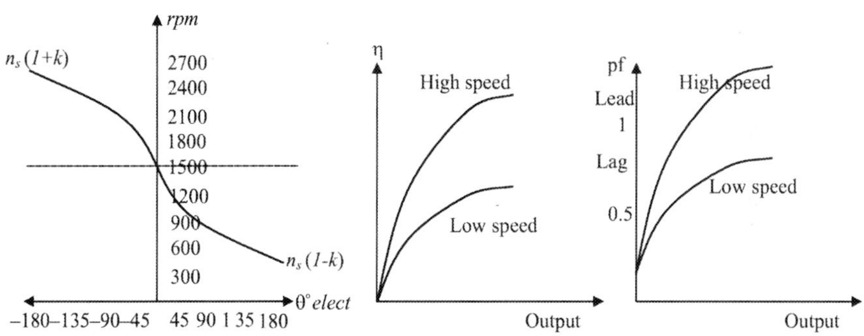

(a) Variation of no load speed with θ (b) Efficiency-output curve (c) PF-output curve

Fig. 9.7. Performance characteristics.

The starting current is limited if injected emf is in phase opposition to E_2. In this case lower speed is obtained.

$$I_{2.st} = \frac{E_2 - E_{j\max} \sin\frac{\theta}{2}}{z_2} \tag{13}$$

Equation 13 shows that the starting current can be limited if the brushes are separated by positive angle $\theta = 180°$, i.e., pole pitch. Thus, no starting devices are required for starting Schrage motor. This maximum brush separation of pole pitch also gives minimum possible speed, n_s $(1 - k)$. To ensure that the brushes are in the minimum speed position at the time of starting, a small switch can be provided and operated by brush gear and interlocked with starting switch.

3.3.2 Advantages of Schrage Motor

The main advantages of the Schrage motor are:

1. wide range of speed control as compared to induction motor
2. better power factor as compared to induction motor
3. more efficiency at higher speed

3.3.3 Disadvantages of Schrage Motor

The disadvantages of the Schrage motor are:

1. due to commutation problem rating is limited to around 25 kW/pole.
2. primary winding on rotor keeps the supply voltage restricted to around 600 V.
3. high cost (less efficiency as compared to same rating induction motor).

3.3.4 Uses of Schrage Motor

It is used as a constant torque drive and power developed varies directly with speed. It is usually applied in printing presses, textile mills, paper mills, conveyors, cranes etc.

4. APPARATUS REQUIRED

1. Schrage machine: Rating: (For example, 5 kW, 440 V. 600 to 2700 rpm)
2. DC Shunt generator: Rating: (For example, 5 kW, 230 V. 1450 rpm)
3. AC Ammeter: one, 0-10 amps.
4. AC Voltmeter: one, 0-300 volts.
5. Wattmeter: one, 300 V, 10 A
6. DC Ammeter: one, 0-10 amps.
7. DC Voltmeter: one, 0-300 volts.
8. Rheostat: one, 1000 ohms, 2 amps.
9. Tachometer or speedometer: one
10. Rheostatic Load: 5 kW, 230 V.

5. CIRCUIT DIAGRAM

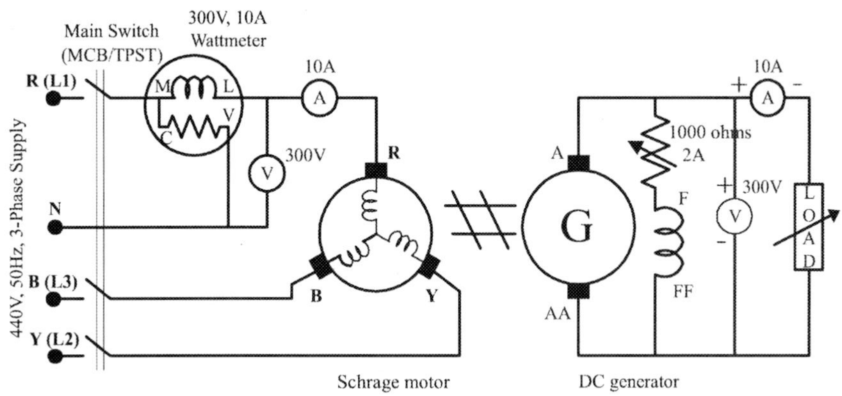

Fig. 9.8. Experimental set-up for studying the variation of speed and load test on Schrage motor.

6. PROCEDURE

Speed Test

1. Make the connections as shown in Fig. 9.8.
2. Put the DC generator field rheostat at maximum resistance point.
3. Keep the brush angle pointer at zero (condition for starting the Schrage motor).
4. Switch on AC supply to Schrage motor by ILTP switch. Press green button of ILTP to start the motor.
5. Note down speed for zero brush angle.
6. Now increase brush angle from zero to 360° gradually at regular intervals and note down the speed for each interval.

Load Test

7. Run the Schrage motor at rated speed of DC shunt generator.
8. Excite the DC shunt generator by decreasing the field rheostat resistance and build up to its rated voltage. Maintain this voltage CONSTANT throughout the experiment.
9. Increase the load gradually and note down the speed and meter readings for each load.

7. OBSERVATION AND CALCULATIONS

Speed Test

Angle θ	Speed N
0	
20	
40	
..	
...	

Load Test

Schrage motor					DC generator			Efficiency of motor
N_m	V_m	I_m	W_m	PF_m, (to be calculated)	I_g	V_g	Output of Gen = $V_g I_g$	η_m
					1A			
.	.				2A			
					3A			
					..			
					...			

8. RESULTS

The following characteristics are drawn:

1. Speed N rpm vs brush angle θ.
2. Speed N rpm vs output of motor.
3. Power factor vs output of motor.
4. Motor load current I_m vs output of motor.
5. Efficiency η_m vs output of motor.

9. CONCLUSION

1. The speed of motor increases as the brush angle is increased. Speed is directly proportional to brush angle. (Rocker and pinion arrangement converts 180° to –180° to zero to 360°.)
2. The speed of motor decreases as the load on motor is increased.
3. The power factor of motor increases or improves as the load on motor is increased.

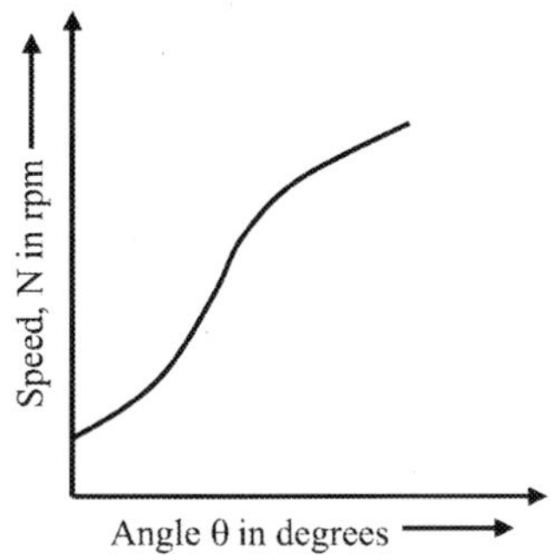

Fig. 9. Variation of no-load speed with brush separation angle.

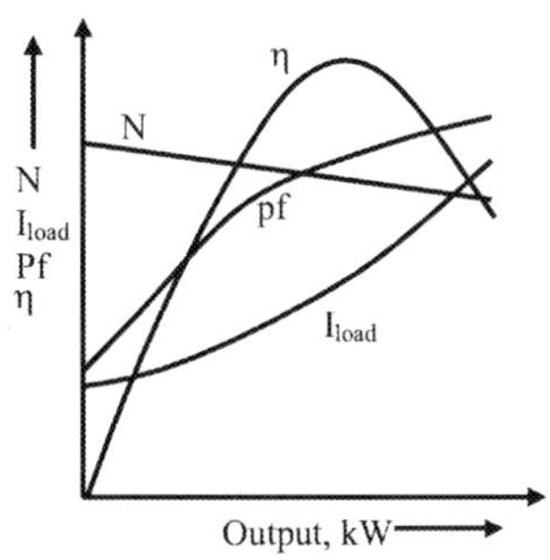

Fig. 10. Performance curves of a Schrage motor under load.

4. The current of motor increases as the load on motor is increased.
5. The efficiency is zero at no load. It increases as the load increases and is maximum when variable losses are equal to constant losses.
6. The efficiency at rated output is less than maximum value and the operating point is after maximum efficiency.

10. POST-EXPERIMENTAL QUIZ

1. Explain the function of tertiary winding.
2. Explain power factor control of Schrage motor.
3. Why the primary winding is housed in the rotor? Explain.

CHAPTER 10

To Plot '*V*' and 'Inverted *V* (Λ)' Curves of a Synchronous Motor

1. OBJECTIVE

The main objectives of this experiment are:

1. to see the effects of excitation on the power factor and the armature current of a synchronous motor.
2. to plot the '*V*' and 'inverted *V*' curves of a synchronous motor.

2. PRE-EXPERIMENTAL QUIZ

1. Why the output power of a synchronous motor remains unchanged due to excitation?
2. What do you mean by nominal excitation in a synchronous motor?
3. Derive the equations for active and reactive power of a synchronous motor.

3. THEORY

3.1 Classification

A synchronous machine is one of the important types of AC electric machines. Depending upon its applications, there are ***three*** forms of this machine.

1. ***Synchronous motor*:** This converts AC electrical power to mechanical power at constant speed.
2. ***Synchronous generator*:** This converts mechanical power to AC electrical power. The generator receives mechanical power from a prime mover, i.e., steam turbine, water turbine, petrol or diesel or gas engine, to which it is mechanically coupled. The generator may be of single-phase, two-phase (90° phase displacement) or three-phase (120° phase displacement). The synchronous generators are also called alternators. The generators, which are driven by steam turbine, are called ***turbo-generators,*** those driven by water turbines are called ***hydro-generators*** and driven by petrol,

diesel or gas engine are called ***engine driven generators.*** The engine driven generators are of small rating, upto 1MW, hydro-generators are of rating upto 100 MW and turbogenerators are of rating upto 1000 MW.

3. ***Synchronous condenser*:** Synchronous motor which operate at leading power factor, is called synchronous condenser and is used to control reactive power in a power system.

3.2 Construction

Similar to other rotating machines, a synchronous machine consists of basic two parts: the stator and the rotor. The stator is the stationary part and the rotor is the rotating part. The stator consists of stator frame, stator core, two end covers, bearing, etc. The rotor core is mounted directly on the shaft or a spider carried by the shaft. The clearance between the stator and the rotor is called *air gap*. The air gap is kept as small as possible to reduce leakage reactance, no load current and to improve the power factor.

3.2.1 Armature

The armature is made up of high-grade alloy ***circular*** steel laminations (cold rolled grain oriented silicon steel sheets) to reduce the eddy current losses. The laminations are *slotted* on the air gap periphery and are insulated from each other. The insulated conductors are placed in these slots. The conductors are connected to form an armature winding which is distributed in nature. The three-phase armature winding may be either star or delta connected. The armature may be on stator or on rotor. Generally, high rating machines have armature on stator.

Armature winding: This is a distributed type winding, which is housed in the armature slots. This winding may be of single-phase, two-phase or three-phase. This winding is similar to the stator winding of induction motor. For large rating machine, 100 MVA, 200 MVA, 500 MVA or more, this winding is on the stator for easy flow of power from or to the stationary armature. Other advantages are: (i) it is economical, (ii) conductors for high rating can be cooled more efficiently and (iii) proper insulation can be provided for higher voltages.

If armature winding is on rotor, then four slip rings are required for making the connections.

When working as a generator, armature winding supplies, delivers or exports AC power to the power system and as a motor, takes, absorbs or imports AC power from the power system.

3.2.2 Field Magnet or Pole Structure

It is made of thin laminations of the same material as that of armature. The magnetic field required for the operation is provided by field magnets similar to those of DC machines and DC supply is needed to excite direct current. The field poles may be on stator or on rotor. Generally, high rating machines have field poles on rotor. The winding wounded on this field pole is field winding.

Field winding: It is a concentrated DC field winding and is similar to the field winding of DC machine. It always absorbs DC power from a DC source for both generator and motor operation. For large rating machine, this winding is on the rotor. The rating of field winding is very less, i.e., 0.5 to 5% of armature rating. As compared to armature, this low power can be easily transferred to rotor with the use of ***two*** slip rings. Other added advantages are: (i) low power field winding gives lighter rotor due to less mass or weight, (ii) less torque is required to rotate the rotor with the use of steam or water, (iii) higher speed and more output are possible for a given dimension.

Damper winding: In addition to the above two windings, sometimes synchronous machines have third type of winding, which is called ***damper winding*** or ammortisseur winding. The damper winding is housed in slots in the field pole faces or pole shoe as shown in Fig. 10.1, similar to compensating winding of DC machines. This winding is made up of copper, brass or aluminium bars. These bars are short circuited by end rings. The winding is similar to the rotor of squirrel cage induction motor. The functions of damper winding are: (i) to start the motor and (ii) to damp the rotor oscillations under disturbed condition.

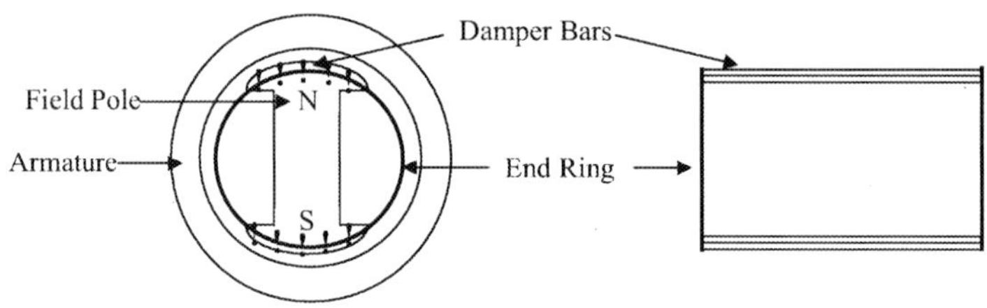

Fig. 10.1. Damper winding.

There are two types of field pole structure:

1. Salient pole or projecting pole type
2. Cylindrical or round or non-salient pole type.

The salient pole type structure on rotor is shown in Fig. 10.2(a) and on stator is shown in Fig. 10.2(b). This has non-uniform air gap and is used for low speed, hydro-generator with number of poles more than four. This type of poles on rotor has small length and large diameter in order to accommodate large number of poles.

The cylindrical pole type structure is shown in Fig. 10.3(a). This has uniform air gap. There are two types of cylindrical ***rotor***, (i) parallel sided slots, as shown in Fig. 10.3(a) and (ii) radial sided slots, as shown in Fig. 10.3(b). This type of rotor is used for high speed, turbo-generator with two or four poles. This is relatively long in length and small in diameter in order to limit the centrifugal forces.

The air gap flux density wave in cylindrical rotor synchronous machine is nearer to sine wave than in salient pole machines.

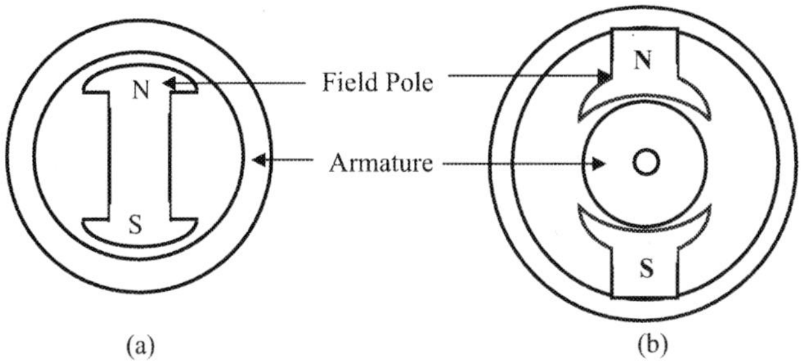

Fig. 10.2. Salient or projecting pole type structure.

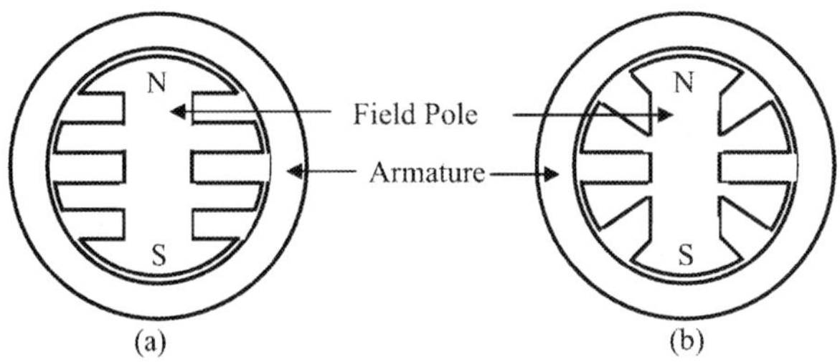

Fig. 10.3. Cylindrical rotor pole type structure.

The cylindrical rotor is the most vital part of a turbogenerator both from mechanical and thermal point of view. In a two-pole generator, the speed is 3000 rpm. The centrifugal forces developed at these speeds create very high mechanical forces and stresses at some part of the rotor and shaft. For this reason, the rotor shaft of a modern turbogenerator is made up of solid steel forging, through the process of heat treatment and machining operation. In large machines chromium–nickel–molybdenum steel with high tensile strength is used.

In case of motor two types of supply are required. One type is AC to the armature winding and second type is DC to the field winding. Therefore motor is a ***doubly excited*** machine.

Horizontal shaft design is usually employed for most of the synchronous generators, motors and synchronous condensers. High power, low speed hydro-generators, installed in low head hydro plants are built with a vertical shaft.

For synchronous machines of large size, adequate cooling arrangements must be incorporated to dissipate the heat produced by various losses. The closed circuit cooling, using air, water, hydrogen or helium, is often used.

3.3 Excitation Systems

The DC supply is required to excite the field winding, in order to produce the magnetic field. In case of generator, alternating voltage is generated in the armature winding due to this magnetic field and rotation. In case of motor, this magnetic field locks with the synchronously rotating AC magnetic field produced by armature winding to have rotation of rotor at synchronous speed. Some of the exciter systems are given below.

3.3.1 DC Exciters

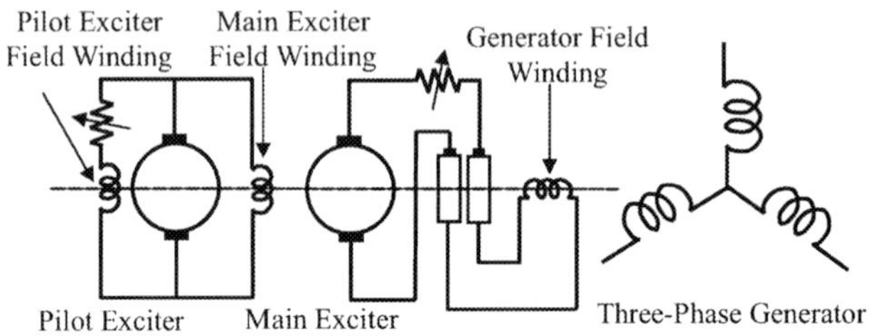

Fig. 10.4. DC exciter.

This is an old conventional method of exciting the field winding. In this method, three machines, namely, pilot exciter, main exciter and the synchronous generator are mechanically coupled and are driven at same speed by the same shaft. The pilot exciter is the DC shunt generator feeding the field winding of main exciter. The main exciter is a separately excited DC generator. The DC output of the main exciter is given to the field winding of the synchronous generator through brushes and slip rings.

3.3.2 Static Excitation

DC excitation is given by battery bank. The three-phase full converter thyristor bridge is used to convert AC to DC. Input to this bridge is given from the generator output. When DC output from thyristor is available, battery bank is bypassed.

Nowadays, for a new generator, where three-phase supply is available from the power system, the static excitation system, without battery bank, is used. The

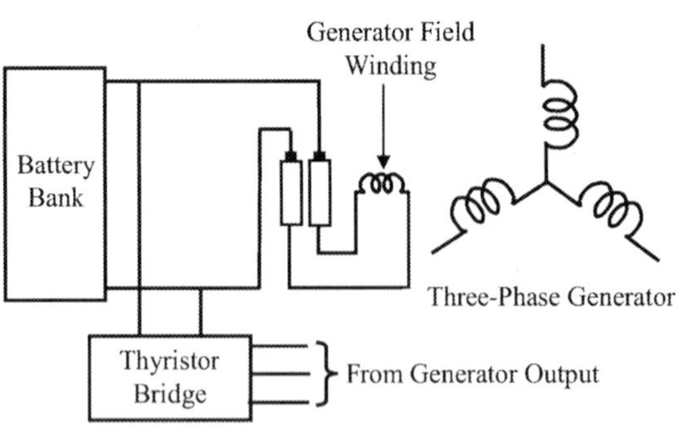

Fig. 10.5. Block diagram of static excitation.

DC output is controlled by controlling the firing angle of the thyristors, which are used in a thyristor bridge.

3.3.3 Brushless Excitation

This method of excitation is shown in Fig. 10.6. In this method, the main shaft of prime mover drives pilot exciter, main exciter and main generator. The diode rectifier bridge is also mounted on the shaft.

The pilot exciter is a three-phase alternator with permanent magnet poles on the rotor and three-phase armature winding on the stator. The three-phase output from pilot exciter is given to thyristor controlled bridge. After rectification DC output is supplied to stationary field winding of main exciter. The three-phase output from armature of main exciter is given to diode rectifier bridge, which is mounted on the same shaft. The DC output of diode rectifier bridge is fed to the field of main generator through hollow shaft without slip rings and brushes. Since in this method, there are no sliding contacts and brushes, it is called brushless excitation.

In laboratory, the field current or excitation is controlled by field rheostat.

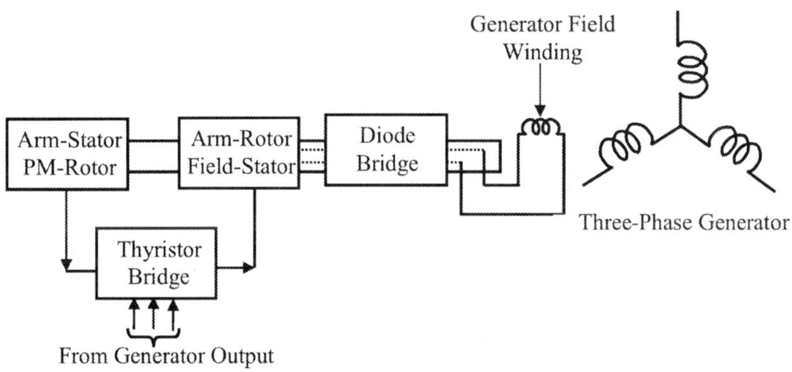

Fig. 10.6. Block diagram of brushless excitation.

3.4 Principle of Operation

Consider armature winding on the stator and field winding on the rotor. In case of synchronous generator, the first rotor is rotated by prime mover at synchronous speed. Then the field winding is excited by DC supply. The field winding is designed in such a way that a sinusoidal flux is set up in the air gap. Due to rotation, this flux links with the armature winding and hence as per Faraday's law sinusoidal voltage is induced in the armature winding.

In case of synchronous motor, first three-phase AC supply is given to the armature winding. Due to this a rotating magnetic field is produced in the air gap. The stator field rotates at synchronous speed, $N_s = 120f/P$. If field winding is excited by DC supply, then steady state magnetic field is set up. When a pair of

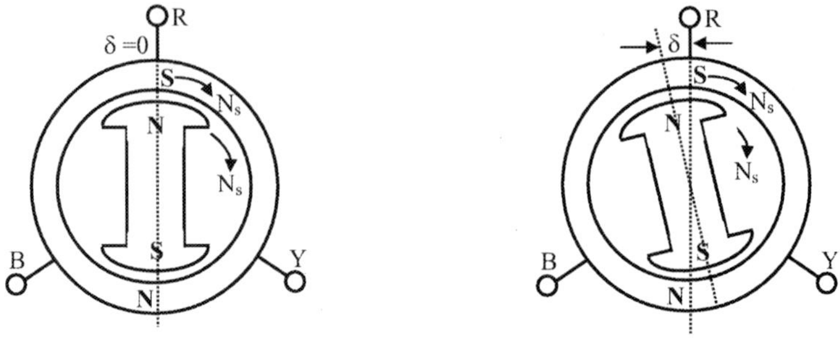

Fig. 10.7. Operating principle.

rotating stator poles sweep across the stationary rotor field poles at synchronous speed, the stator N pole will tend to rotate the rotor in one direction and after 10 milliseconds the stator S pole will tend to rotate the rotor in reverse direction. However due to the rotor inertia, rotor cannot follow the stator poles at the synchronous speed. As a result, the rotor does not move or rotate and starting torque is zero. Therefore synchronous motor is ***not a self-starting motor***.

In order to develop a continuous torque, the two fields must be stationary with respect to each other. This is possible when the rotor also rotates at synchronous speed. The rotor is rotated at synchronous speed by prime mover, i.e., by any AC or DC motor. At that speed there is magnetic locking between stator N pole with rotor S pole and stator S pole with rotor N pole. The relative speed between the two fields is zero and rotor will now experience a torque. Thus, the rotating magnetic field of armature "***drags***" the DC field poles or rotor poles.

If prime mover of rotor is removed, then rotor will continue to rotate in the same direction as that of rotating magnetic field, at synchronous speed. Hence, the name, synchronous motor.

If some mechanical load is applied to the shaft of the motor, rotor lags the stator rotating magnetic field by some angle δ, which is called torque angle or load angle or power angle. With this lagging angle, rotor rotates at synchronous speed due to magnetic locking between stator and rotor poles.

3.5 Starting

We have seen that synchronous motor is not a self starting motor. It can be started by the following methods:

1. Auxiliary motor starting
2. Damper winding starting (SCIM)
3. SRIM starting with high torque

3.5.1 Auxiliary Motor Starting

The auxiliary motor may be AC motor or DC motor. This motor is mechanically coupled with the synchronous motor. The direction of rotation of rotor of

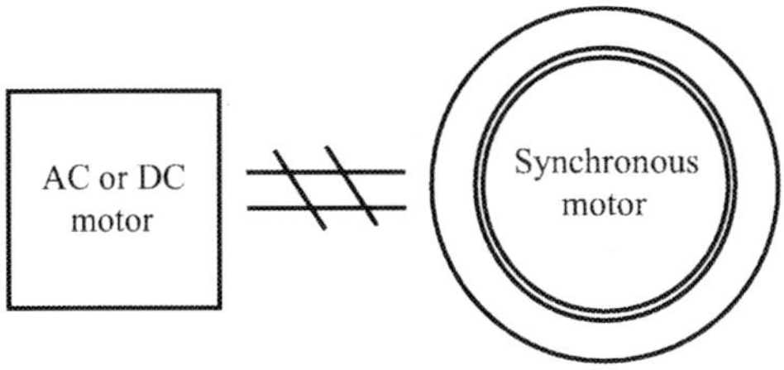

Fig. 10.8. Auxiliary motor starting.

synchronous motor can be checked by giving reduced voltage to the armature with the use of three-phase autotransformer. For this the field winding can be shorted with external resistance for small interval of time. Then auxiliary motor is started and run near to the synchronous speed in the assigned direction of rotation of rotor of synchronous motor. The armature winding of synchronous motor is energized from three-phase supply. Then DC supply is given to the field winding of synchronous motor. Due to this field poles get locked with those produced by armature winding and the main motor starts running at synchronous motor at synchronous speed. The auxiliary motor can now be decoupled or disconnected from its supply.

3.5.2 Damper Winding Starting

Today the most widely used method of starting a synchronous motor is to use damper windings. When a 3-phase supply is given to the armature, a rotating magnetic field is set up. This rotating field and damper winding develop induction motor torque. Therefore rotor is accelerated upto synchronous speed. The DC excitation is applied to the field winding. The rotor will then pull into step with the stator magnetic field.

3.5.3 SRIM Starting with High Torque

Auxiliary motor starting and damper winding starting are used when the load requires low rotating torque. For load requiring high starting torques, this method is used.

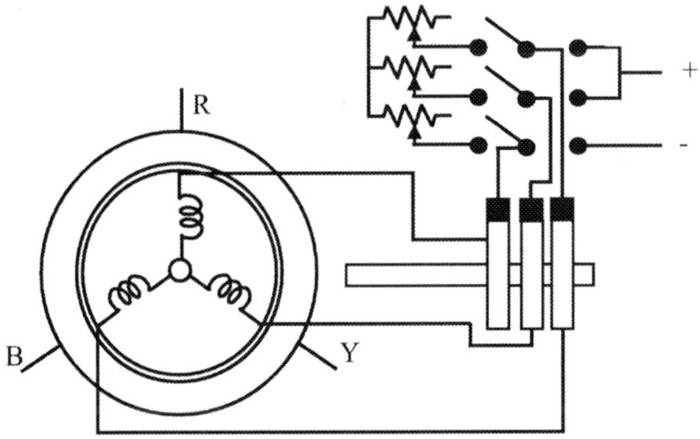

Fig. 10.9. SRIM starting with high torque.

The rotor of synchronous motor is similar to the rotor of wound rotor induction motor. Initially the motor is started as SRIM and then synchronous motor, therefore this type of synchronous motor may be called syn-duction motor.

At the time of starting, additional high resistance is inserted in the rotor circuit in order to increase the starting torque. This is obtained by switching on the triple pole double throw switch toward resistance side. As the speed increases, the additional rotor resistance is gradually reduced to zero. With this zero resistance, motor runs at a speed near to synchronous speed. At this time the triple pole double throw switch is connected to DC supply. The rotor poles are created and are attracted by stator poles. Thus, synchronism is achieved.

3.6 Phasor Diagram

A synchronous motor is same in all respects as a synchronous generator except that the direction of power flow is reversed. For simplicity, a cylindrical rotor synchronous generator is considered. Initially the generator is run at synchronous speed by adjusting the speed of its prime mover. The field winding is now excited by DC source. The no load terminal voltage or armature voltage E_f is generated and is made equal to rated value by adjusting field current. The per phase generated emf is given by

$$E_f = 4.44\, f\, \Phi_f\, N_{ph}\, K_w \tag{1}$$

where E_f = no load voltage or excitation voltage or excitation emf,
f = frequency of generated emf, $PN_s/120$,
Φ_f = flux per pole produced by DC current in the field winding,
N_{ph} = series turns per phase,
K_w = winding factor, varies from 0.9 to 0.98.

The field mmf per pole F_f is equal to $I_f N_f$. As saturation is neglected, the field mmf F_f and field flux Φ_f can be drawn in phase in phasor diagram, as shown in Fig. 10.10. As the field mmf F_f and field flux Φ_f are sinusoidally distributed along the air gap periphery, they can be represented by phasors. The generated emf lags the field flux by 90°.

When alternator is connected to three-phase load, three-phase generated emf in armature will give rise to three-phase balanced current. Consider a general case of lagging power factor, with the position of armature current I_a phasor as shown. This armature current sets up armature mmf F_a, which is called armature reaction mmf. Due to F_a, the armature flux Φ_a is set up. The relative speed between F_f and F_a is zero. The phasor sum of F_f and F_a gives the resultant or air gap mmf F_r. Mathematically, it can be written as $\overline{F}_r = \overline{F}_f + \overline{F}_a$.

For uniform air gap, the reluctance is constant at all angular position of the rotor. Therefore, the resultant or air gap flux Φ_r is obtained by phasor sum of Φ_f and Φ_a. The space phasors and time phasor quantities, as usual, are taken to rotate conterclockwise and this is shown in generator phasor diagram by an arrow marked ω. The field mmf wave or field poles have always a tendency to align themselves

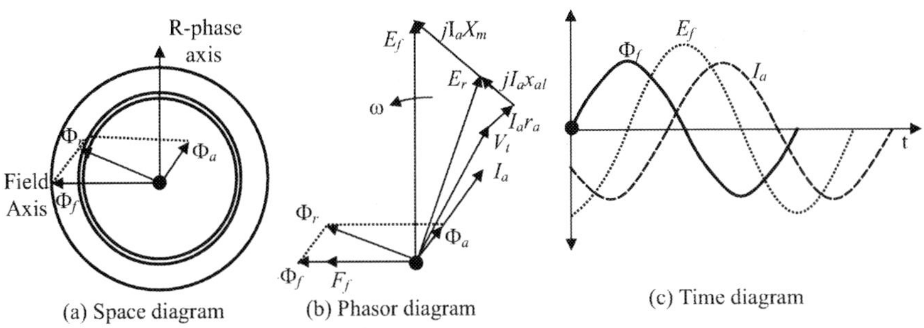

Fig. 10.10. Representation of flux, MMF, EMF and current.

along the air gap flux Φ_r. This resulted in clockwise rotation due to electromagnetic torque T_e. Since this torque opposes the direction of rotation, ω, the machine must act as a generator. It can also be stated that for generator, the field poles must be driven ahead of the resultant air gap flux by the forward torque of the prime mover.

This air gap flux induces emf E_r lagging by 90° to it. This E_r is called ***air gap voltage*** or ***voltage behind leakage impedance***. Deducting from E_r, the armature resistance drop I_ar_a and leakage reactance drop jI_ax_{al}, the terminal voltage V_t or phase voltage of armature winding is obtained. If mutual reactance drop jI_aX_m is added to E_r, then the excitation voltage E_f is obtained. Therefore, the voltage equation for the generator, can be written as

$$\overline{E}_f = \overline{V}_t + \overline{I}_a(r_a + jx_{al} + jX_m)$$

$$\overline{E}_f = \overline{V}_t + \overline{I}_a(r_a + jX_s) = \overline{V}_t + \overline{I}_aZ_s \qquad (2)$$

where X_s = synchronous reactance and Z_s = synchronous impedance.

The angle between V_t and I_a is power factor angle θ. The angle between E_f and I_a is internal power factor angle. The angle between E_f and V_t is power angle δ. The angle between Φ_f and Φ_a is equal to 90 + ψ and for the operation of either motor or generator is angle is always less than 180°. In generator operation, since excitation is given first and then there is building up of terminal voltage, E_f leads the terminal voltage V_t by an angle δ.

For a cylindrical rotor synchronous motor, the flow of power and current directions are reversed. In equation 2, put $I_a = -I_a$.

$$\therefore \overline{E}_f = \overline{V}_t - \overline{I}_a(r_a + jX_s) = \overline{V}_t - \overline{I}_aZ_s$$

$$\overline{V}_t = \overline{E}_f + \overline{I}_a(r_a + jX_s) = \overline{E}_f + \overline{I}_aZ_s \qquad (3)$$

For motor operation, first terminal voltage is given to the armature and after rotating the field rotor near to synchronous speed, excitation is given. Therefore, V_t leads the terminal voltage E_f by an angle δ.

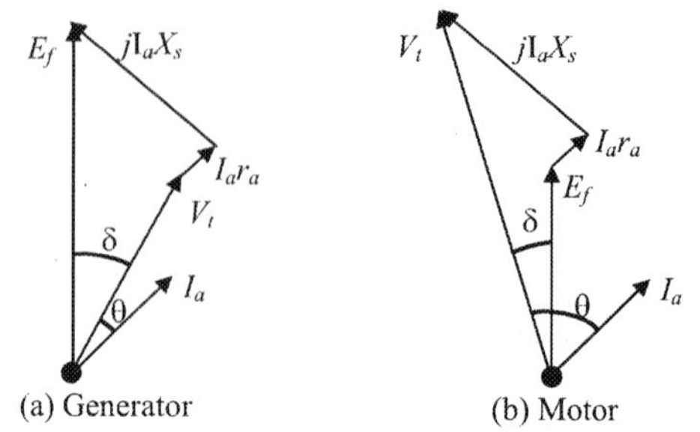

Fig. 10.11. Phasor diagrams of cylindrical rotor synchronous machine.

The phasor diagrams for cylindrical rotor synchronous generator and motor, as per equations 2 and 3 are drawn in Fig 10.11.

For salient pole synchronous machines, there are two reactances due to non-uniform air gap. The reactance along field pole axis or direct axis (d-axis) is called direct axis synchronous reactance, X_d. The axis, 90° away from d-axis is called quadrature axis (q-axis) and the reactance along this axis is called quadrature axis synchronous reactance, X_d. Therefore, by using equation 2, the voltage equation for salient pole synchronous generator can be written as

$$\overline{E}_f = \overline{V}_t + \overline{I}_a r_a + j\overline{I}_d X_d + j\overline{I}_q X_q \tag{4}$$

where I_d = component of armature current along d-axis,

I_q = component of armature current along q-axis.

Similarly, the voltage equation for salient pole synchronous motor can be written as

$$\overline{V}_t = \overline{E}_f + \overline{I}_a r_a + j\overline{I}_d X_d + j\overline{I}_q X_q \tag{5}$$

The phasor diagrams for salient pole synchronous generator and motor, as per the equations 4 and 5 are drawn in Fig 10.12.

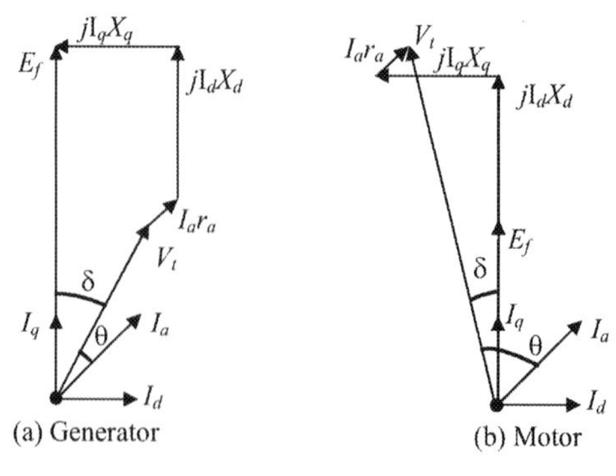

Fig. 10.12. Phasor diagrams of salient pole synchronous machine.

3.7 Active Power

An infinite bus has a constant voltage and frequency. A synchronous generator, if connected to an infinite bus supplies electrical power at constant voltage. The excitation voltage, E_f can be varied by varying the field current while the electrical power, P_e can be controlled by controlling the mechanical power of the prime mover.

If core loss and mechanical loss, i.e., friction and windage loss are subtracted from the mechanical power input at the shaft, P_m is obtained. If armature copper loss is subtracted from P_m, then electrical power P_e is obtained. Therefore,

P_m = power being converted from mechanical form to electrical form or
= equivalent of mechanical power converted to electrical power
= mechanical power input at the shaft-constant loss
= P_e + armature copper loss

3.7.1 Cylindrical Rotor Synchronous Generator

$$P_e = 3V_t I_a \cos\theta$$

With the use of dot product, this can be written as

$$P_e = 3V_t \bullet I_a$$

$$P_m = 3(V_t I_a \cos\theta + I_a^2 r_a)$$

$$= 3(V_t \bullet I_a + [I_a Z_s] \bullet I_a) \text{ (since } Z_s \cos\theta_z = r_a)$$

$$= 3([V_t + I_a Z_s] \bullet I_a)$$

$$= 3E_f \bullet I_a \quad (6)$$

= electromagnetic power

Thus, electrical power P_e is the dot product of V_t and I_a and electromagnetic power P_m is the dot product of E_f and I_a. Therefore, in order to derive the expression for electrical power, V_t is taken as a reference phasor and for electromagnetic power, E_f is taken as a reference phasor. From Fig. 10.11(a), for P_e, the I_a can be written as

$$I_a = \frac{E_f\angle\delta - V_t\angle 0}{Z_s\angle\theta_z}$$

$$\therefore\ P_e = 3V_t \bullet \frac{E_f\angle\delta - V_t\angle 0}{Z_s\angle\theta_z} \quad (V_t \text{ as a reference phasor})$$

$$= 3\left\{\frac{V_t E_f}{Z_s}\angle\delta - \theta_z - \frac{V_t^2}{Z_s}\angle -\theta_z\right\}$$

$$= 3\left\{\frac{V_t E_f}{Z_s}\cos(\delta - \theta_z) - \frac{V_t^2}{Z_s}\cos(-\theta_z)\right\}$$

$$= 3\left\{\frac{V_t E_f}{Z_s}\cos(\theta_z - \delta) - \frac{V_t^2}{Z_s}\cos\theta_z\right\} \quad (7)$$

For maximum value of P_e, $\delta = \theta_z$ (= 80 to 85°), i.e., power angle should be equal to armature impedance angle. If resistance is negligible, then $\theta_z = 90$ and $X_s = X_s$, therefore, electrical power is

$$P_e = 3\frac{V_t E_f}{Z_s}\sin\delta \quad (8)$$

For maximum value of P_e, $\delta = 90°$. The power angle characteristics can be drawn from equation 8. The same expression is applicable for cylindrical rotor synchronous motor.

With the reference phasor E_f, for P_m, from Fig. 10.11(a), the I_a can be written as

$$I_a = \frac{E_f \angle 0 - V_t \angle -\delta}{Z_s \angle \theta_z}$$

$$P_m = 3E_f \bullet \frac{E_f \angle 0 - V_t \angle -\delta}{Z_s \angle \theta_z}$$

$$P_m = 3\left\{\frac{E_f^2}{Z_s}\angle\theta_z - \frac{E_f V_t \angle -\delta - \theta_z}{Z_s}\right\}$$

$$= 3\left\{\frac{E_f^2}{Z_s}\cos\theta_z - \frac{E_f V_t}{Z_s}\cos(\delta + \theta_z)\right\} \quad (9)$$

For maximum value of P_m, $\delta + \theta_z = 180$ or $\delta = 180 - \theta_z$ (= 95 to 100). If resistance is negligible, then $\theta_z = 90$ and $X_s = X_s$, therefore, electromagnetic power is

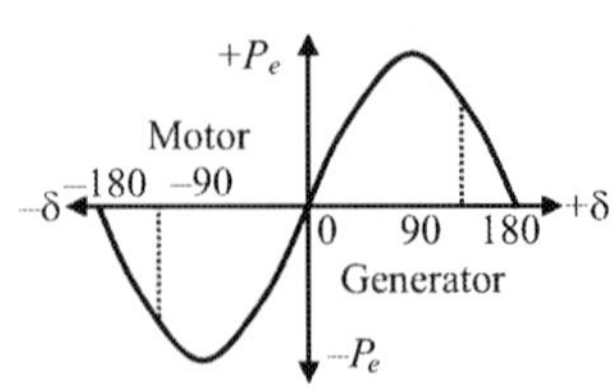

Fig. 10.13. Power-angle characteristic.

$$P_m = 3\frac{E_f V_t}{Z_s}\sin\delta \tag{10}$$

For maximum value of P_m, $\delta = 90°$. The power angle characteristics can be drawn from equation 10, as shown in Fig. 10.13. The same expression is applicable for cylindrical rotor synchronous motor.

It is clear that when angle is positive P_m and P_e are positive and electrical power is delivered to the power system, i.e., generator operation. When angle is negative P_m and P_e are negative and electrical power is absorbed from the power system, i.e., motor operation.

As seen from equation 8 and 10, that if resistance is neglected, electromagnetic power and electrical power are same.

3.7.2 Salient Pole Synchronous Generator

For simplicity neglect the resistance. Referring to Fig. 10.12(a), the power expression can be written as

$$P_m = P_e = 3V_t I_a \cos\theta$$

= voltage component x in phase current component

$$= 3\{V_t I_q \cos\delta + V_t I_d \cos(90 - \delta)\} = 3\{V_t I_q \cos\delta + V_t I_d \sin\delta\} \tag{11}$$

Now,

$$X_d I_d = E_f - V_t \cos\delta \text{ and } X_q I_q = V_t \sin\delta$$

$$\therefore I_d = \frac{E_f - V_t\cos\delta}{X_d} \text{ and } I_q = \frac{V_t\sin\delta}{X_q} \tag{12}$$

Putting equation 11 in equation 12, we get

$$P_m = P_e = 3\left\{\frac{V_t^2\sin\delta\cos\delta}{X_q} + \frac{E_f V_t\sin\delta}{X_d} - \frac{V_t^2\sin\delta\cos\delta}{X_d}\right\}$$

$$= 3\left\{\frac{E_f V_t\sin\delta}{X_d} + \frac{V_t^2}{2}\left(\frac{1}{X_q} - \frac{1}{X_d}\right)\sin 2\delta\right\} \tag{13}$$

The power angle characteristics can be drawn by using equation 13, as shown in Fig. 10.14. The power is maximum when an angle δ is from 80 to 85° approximately. The same expression is applicable for salient pole synchronous motor. The first part of the equation 13, with fundamental component of δ, is called electromagnetic power and the second part with second harmonic component of δ, is called reluctance power. The reluctance power is due to saliency.

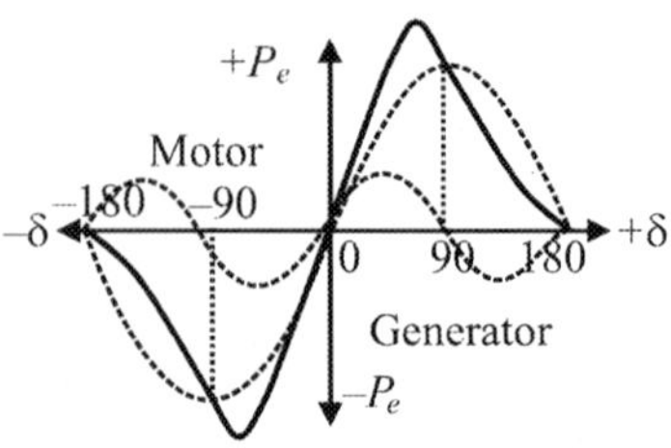

Fig. 10.14. Power-angle characteristic for salient pole machine.

3.8 Reactive Power Q

The reactive power Q is written as

$$Q = 3\ V_t I_a \sin\theta$$

= Voltage component × Quadrature lagging current component

Consider salient pole synchronous generator. For simplicity neglect the resistance and referring Fig. 10.12(a), the reactive power can be written as

$$Q = 3\{V_t I_q \sin\delta - V_t I_d \sin(90 - \delta)\} = 3\{V_t I_q \sin\delta - V_t I_d \cos\delta\} \tag{14}$$

Putting equation 11 in equation 14, for I_d and I_q, we get

$$Q = 3\left\{\frac{V_t^2 \sin^2\delta}{X_q} - \frac{E_f V_t \cos\delta}{X_d} + \frac{V_t^2 \cos^2\delta}{X_d}\right\}$$

$$= 3\left\{\frac{V_t^2 \sin^2\delta}{X_q} - \frac{E_f V_t \cos\delta}{X_d} + \frac{V_t^2}{X_d} - \frac{V_t^2 \sin^2\delta}{X_d}\right\}$$

$$= 3\left\{\frac{V_t^2}{X_d} + V_t^2\left(\frac{1}{X_q} - \frac{1}{X_d}\right)\sin^2\delta - \frac{E_f V_t}{X_d}\cos\delta\right\} \tag{15}$$

For cylindrical rotor synchronous generator, $X_d = X_q = X_s$.

$$\therefore Q = 3\left\{\frac{V_t}{X_s}\left(V_t - E_f \cos\delta\right)\right\} \tag{16}$$

Equation 16 shows that when $V_t = E_f \cos\delta$, i.e., under normal excitation, $Q = 0$, and generator operates at unity power factor. In case $E_f \cos\delta < V_t$, the generator is under-excited, ***power factor is leading***, Q is positive and generator ***delivers reactive power*** to the power system. In case $E_f \cos\delta > V_t$, the generator is over-excited, ***power factor is lagging***, Q is negative and generator ***absorbs reactive power*** from the power system.

Equation 16 is also applicable for motor. When $V_t = E_f \cos\delta$, i.e., under normal excitation, $Q = 0$, and motor operates at unity power factor. In case $E_f \cos\delta < V_t$, the motor is under-excited, ***power factor is lagging***, Q is positive and motor ***absorbs reactive power*** from the power system. In case $E_f \cos \delta > V_t$, the motor is over-excited, ***power factor is leading***, Q is negative and motor ***delivers reactive power*** from the power system.

Thus, synchronous machine can be operated at lagging as well as leading power factor. The over-excited synchronous machine delivers reactive power to the power system and under-excited synchronous machine absorbs reactive power from the power system.

3.9 Effect of Varying Excitation

In a synchronous motor, V and inverted V (Λ) curves are obtained by the variation of field excitation. The field current or excitation of a synchronous motor is controlled with the help of field rheostat. Consider the operation of a cylindrical rotor synchronous motor, for which the general voltage equation is $\overline{V}_t = \overline{E}_f + \overline{I}_a Z_s$. For motor the phasor V_t leads the phasor E_f, since the voltage V_t is given first to the armature winding, the rotor is run near to synchronous speed by prime mover and then field winding which is on the rotor is excited by E_f or DC voltage, for the operation of motor.

If armature resistance is neglected, then the above general voltage equation becomes $\overline{V}_t = \overline{E}_f + j\overline{I}_a X_s$.

The electrical power received from the source of excitation, i.e., DC source can be utilized only for production of the field flux by the electromagnets and then the power is wasted as I^2R loss in the field winding. Thus it cannot contribute to the shaft power, i.e., mechanical output of the motor. On the other hand, the three-phase AC armature currents are responsible for producing rotating magnetic field, the torque, and then the mechanical power. Thus three-phase AC source must feed an active component of current to the motor. For the constant mechanical output, the active component or active power remains constant. Therefore, the variation of field excitation does not result into the change in active power but the power factor.

The effect of varying the field excitation is considered first at no load and then on load.

3.9.1 No-Load Operation

Since all losses are neglected, load angle δ, between E_f and V_t, at no load is zero.

In view of this, power transferred $\left\{ P = \frac{E_f V_t}{X_s} \sin \delta \right\}$ either to, or from the bus, is zero. If excitation voltage E_f is equal to the bus voltage V_t, then at no load, no current flows, either to, or from the bus. Under this condition of no load current flow, the synchronous motor is said to be ***floating*** on the bus.

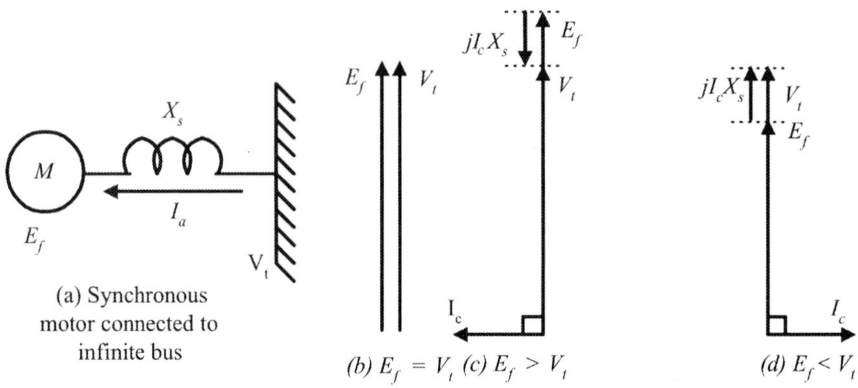

Fig. 10.15. Effect of changing excitation of unloaded synchronous motor.

In case motor at no load is ***over-excited, E_f becomes more than V_t*** and a circulating current $\left\{\bar{I}_c = \dfrac{\bar{E}_f - \bar{V}_t}{jX_s}\right\}$ is set up. Load angle δ remains zero, therefore current I_c ***leads*** E_f and V_t by 90° as shown in Fig. 10.13(c). Since I_c leads E_f by 90°, it produces demagnetizing armature mmf which nullifies the effect of increased field mmf. The motor ***delivers reactive power*** to the bus because of $E_f > V_t$ and it operates as a ***condenser***.

Now if the motor is ***under-excited, E_f becomes less than V_t*** and a circulating current I_c is set up but this time current I_c ***lags*** E_f and V_t by 90° as shown in Fig. 10.13(d). Since I_c lags E_f by 90°, it produces magnetizing armature mmf which adds to the decreased field mmf, thereby making up for the loss of field ampere turns. The motor ***takes reactive power*** from the bus because of $E_f < V_t$.

In both the cases above, no active power is delivered because I_c is in quadrature with V_t (V_tI_c cos 90 = 0) and also load angle δ is zero.

If all the losses are considered, small δ appears between E_f and V_t, and small active power will be absorbed by the motor. The field current required for setting up the air gap flux, as demanded by constant V_t, for small δ, is called ***normal*** excitation. The motor operates at ***unity power factor*** and armature current I_a is ***minimum***. The reactive power is zero.

If the field current is made more than its normal excitation, i.e., the motor is over-excited, and motor operates at leading power factor. If the field current is made less than its normal excitation, i.e., the motor is under-excited, and motor operates at lagging power factor.

3.9.2 On-Load Operation

Now suppose that the motor is loaded with excitation voltage E_f, power angle δ and **unity power factor** with **minimum current I_a** as illustrated in Fig. 10.16. If

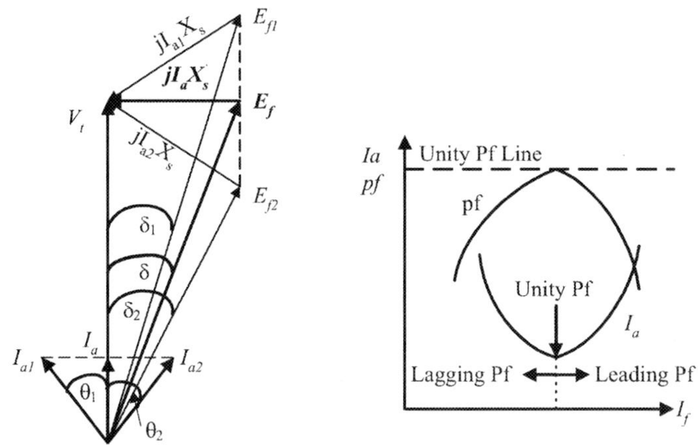

(a) Phasor diagram (b) V and inverted V curves

Fig. 10.16. Effect of changing field excitation of a loaded synchronous motor.

mechanical load is kept constant, then for negligible losses, $P = \dfrac{E_f V_t}{X_s} \sin\delta$ must remain constant.

An increase of E_f to E_{f1} (***over-excitation***) causes the load angle to decrease to δ_1, which can be obtained from the relation $E_{f1} \sin \delta_1 = E_f \sin \delta$ because V_t and X_s are constant. The synchronous reactance drop increases and this has the effect of increasing armature current I_a to I_{a1}. The examination of Fig. 10.16 (a) reveals that the increased current I_{a1} has a real component I_a and a demagnetizing component I_{c1}, which results in ***leading power factor*** $\cos \theta_{2-}$ and motor delivers reactive power to the bus.

If E_f is decreased to E_{f2} (***under-excitation***), the load angle increases to δ_2, which can be obtained from the relation $E_{f2} \sin \delta_2 = E_f \sin \delta$ because V_t and X_s are constant. The synchronous reactance drop increases and this has the effect of increasing armature current I_a to I_{a2}. The examination of Fig. 10.16 (a) reveals that the increased current I_{a2} has a real component I_a and a magnetizing component I_{c2}, which results in ***lagging power factor*** $\cos \theta_{2-}$ and motor absorbs reactive power from the bus.

Note that the real components of I_{a1} and I_{a2} are equal to I_a, i.e. $I_{a1} \cos \theta_1 = I_{a2} \cos \theta_2 = I_a$ and as a result of it, $V_t I_{a1} \cos \theta_1 = V_t I_{a2} \cos \theta_2 = V_t I_a$ which means that the active power of the motor remains unchanged. Thus, the adjustment of field rheostat, for changing the excitation, has no effect on the power absorbed by the motor from the bus. However, a change in the motor field current alters its operating power factor as shown in Fig. 10.16(b).

If the variation of armature current with the change in field current is plotted, then the curve resembles to the letter *V*. This is called ***V curve*** of a synchronous motor. If motor power factor is plotted as a function of the field current, then this curve resembles to *inverted* V(Λ). This is called ***inverted V*** (Λ) ***curve*** of a synchronous motor.

4. APPARATUS REQUIRED

1. Synchronous motor: Rating: Armature: For example, 7.5 kVA, 440 V, 1500 rpm Field: 20 V, 20 A
2. DC Shunt Generator: Rating: 5 kW, 230 V, 1440 rpm
3. AC Ammeter: one, 0-15 amps.
4. Three-phase power factor meter: one, 400 V, 15 A.
5. DC ammeter: two, 0-30 amps, & 0-10 amps.
6. DC voltmeter: one, 0-300 volts.
7. Rheostat: two, 5.6 ohms, 20 amps & 370 ohms, 2 amps.
8. Rheostatic load: 5 kW, 230 V

5. CIRCUIT DIAGRAM

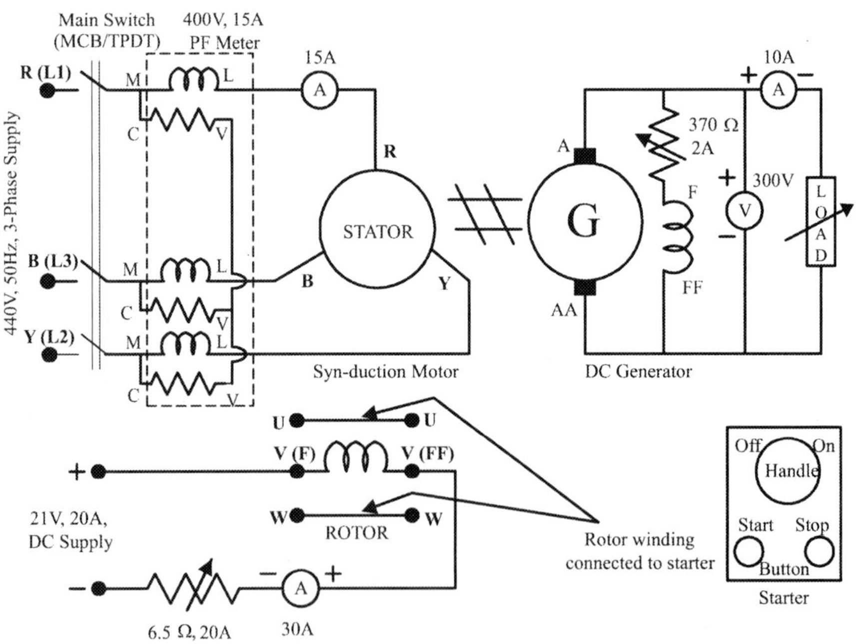

Fig. 10.17. Experimental set-up for plotting *V* and inverted *V* curves of a synchronous motor.

6. PROCEDURE

1. Make the connections as shown in Fig. 10.17.
2. Put the DC generator field rheostat at maximum resistance point.

3. Synchronous motor is started by induction motor starting. Increase the rotor resistance to have more starting torque. This is obtained by moving the handle of starter towards OFF position.
4. Switch on AC supply to synchronous motor by main switch. Press start button of the starter to start the motor.
5. Decrease the rotor resistance by moving handle of starter towards ON position. The speed of motor increases to rated value.
6. Switch on DC supply of 20 V to rotor or field circuit of synchronous motor. With some field current, the magnetic locking (between stator and rotor) sound can be heard.
7. For no load on synchronous motor, keep the field winding of DC generator unexcited.
8. With no load, stepwise increase the field current of synchronous motor and note down the armature current and power factor. Cover a range from lagging pf to low leading pf through unity pf. Note down the reading of armature current and field current at unity power factor.
9. Excite the DC shunt generator to generate rated voltage. Increase the load on DC generator to same value (say, 5 A).
10. With this load of 5 A, stepwise decrease the field current of synchronous motor and note down the armature current and power factor. Cover a range from low leading pf to lagging pf through unity pf. Note that armature current is minimum when the power factor is unity.

7. OBSERVATION

Set-I: No load condition: $I_{G(DC)}$ = Zero amp

Sr. No.	Field Current I_f A	Power factor	Armature Current I_a A
1		lag	
2		lag	
3		**1.0**	**Min**
..		lead	
...		lead	

Set-II: On load condition: V_G = 220 V, $I_{G(DC)}$ = 5A prepare similar table.

8. RESULT

The variation of armature current and its power factor with respect to field current are drawn as shown in Fig. 10.16(b) and studied.

9. CONCLUSION

1. The curve of armature current Vs field current is V curve.
2. The curve of power factor Vs field current is inverted V (^) curve.
3. The armature current is minimum when the power factor is unity.
4. Under-excited ($E_f < V_b$) synchronous motor operates at lagging power factor and absorbs reactive power.
5. Over-excited ($E_f > V_b$) synchronous motor operates at leading power factor and supplies reactive power.
6. Active power remains unchanged.

10. POST-EXPERIMENTAL QUIZ

1. Draw the phasor diagrams for synchronous motor and synchronous generator.
2. If synchronous motor is started by induction motor starting method, then explain why starter resistance is kept at maximum value at the time of start?
3. Explain the over-excited operation of a synchronous motor.

CHAPTER 11

To Determine the Ratio X_q/X_d for Three-Phase Alternator by using "Slip Test"

1. OBJECTIVE

The main objectives of this experiment are:

1. to determine the values of d-axis and q-axis synchronous reactances.
2. to determine the ratio X_q/X_d for synchronous machine.

2. PRE-EXPERIMENTAL QUIZ

1. What do you mean by direct axis?
2. Define X_d and X_q of synchronous machine.
3. Explain armature leakage reactance of synchronous machine.

3. THEORY

3.1 Two-Reaction Theory

In a cylindrical rotor synchronous machine, air gap is uniform. Therefore its reactance remains the same irrespective of the spatial position of the rotor. The effect of armature reaction, fluxes and voltages induced can be treated in a simple way with the concept of a synchronous reactance and taking it as constant for all position of field poles with respect to armature. On the other hand, in a salient pole synchronous machine, air gap is non-uniform, due to which the reluctance and reactance vary with the rotor position. Therefore, the cylindrical rotor theory, based on constant air gap reluctance, cannot be applied to a salient pole machine which has different air gap reluctances along its periphery.

The salient pole synchronous machine can be represented as shown in Fig. 11.1. The armature consists of three phase windings, if connected to balanced three-phase supply; a rotating magnetic field is produced. The speed of this rotating field is synchronous speed. The field consists of DC field winding, if excited by DC supply; poles are created on the rotor. If rotor is rotated at synchronous speed by prime mover or motor, the resultant armature mmf is stationary with respect to field mmf.

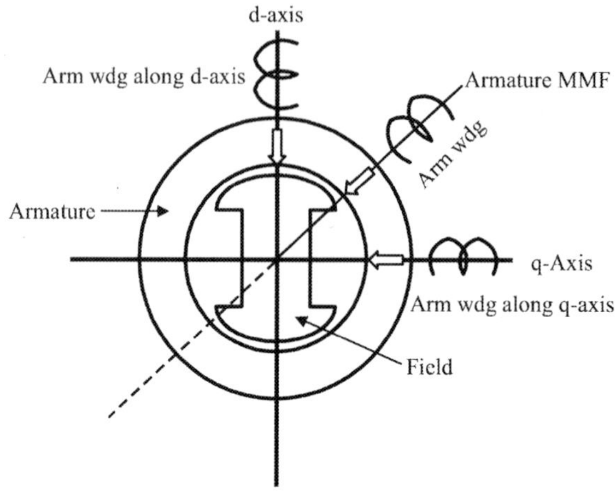

Fig. 11.1. Armature MMF resolved into its D and Q components.

The effect of armature mmf cannot be calculated unless it is resolved into two components (interpolar axis).

These are taken along:

(a) The pole axis, known as **direct-axis** (d-axis) (polar axis) and

(b) The axis quadrature to this, i.e., 90° electrical lagging to this, known as **quadrature axis** (q-axis) (interpolar axis).

As per two reaction theory, the sinusoidal armature mmf F_a is resolved into two sinusoidal components, one F_{ad} along d-axis and the other F_{aq} along q-axis. It is seen that F_{ad} experiences minimum air gap and reluctance along d-axis and F_{aq} experiences maximum air gap and reluctance along q-axis. The mmf F_a, F_{ad}, and F_{aq} are produced due to armature current I_a, d-axis current I_d and q-axis current I_q respectively. Similarly, the armature flux Φ_a is resolved into two components, one Φ_{ad} along d-axis and the other Φ_{aq} along q-axis. The Φ_{ad} is having maximum value due to minimum reluctance and the Φ_{aq} is having minimum value due to maximum reluctance. Thus, depending on flux linkage per armature ampere, there are two reactances, one X_d, d-axis synchronous reactance along direct axis and other X_q, q-axis synchronous reactance along quadrature axis.

Due to F_{ad}, there is a induced voltage E_{ad} lagging by 90° to F_{ad} and due to F_{aq}, there is a induced voltage E_{aq} lagging by 90° to F_{aq}. In terms of reactances and currents these voltages can be written as

$$\overline{E}_{ad} = j\overline{I}_d X_{md} \text{ and } \overline{E}_{aq} = j\overline{I}_q X_{mq}$$

Here X_{md} and X_{mq} are the d-axis and q-axis magnetizing reactances (or armature reaction reactances) of a salient pole synchronous machine.

Similar to cylindrical rotor synchronous generator, if E_{ad} and E_{aq} are subtracted from no load voltage E_f, then air gap voltage E_r is obtained.

From E_r if armature resistance drop $I_a r_a$ and leakage reactance drops $I_d x_{al}$ and $I_q x_{al}$ are subtracted, then terminal voltage V_t is obtained. Therefore, the voltage equation for salient pole synchronous generator is

$$\overline{E}_f = \overline{V}_t + \overline{I}_a r_a + j\overline{I}_d x_{al} + j\overline{I}_q x_{al} + j\overline{I}_d X_{md} + j\overline{I}_q X_{mq}$$

$$\overline{E}_f = \overline{V}_t + \overline{I}_a r_a + j\overline{I}_d (x_{al} + X_{md}) + j\overline{I}_q (x_{al} + X_{mq})$$

$$\overline{E}_f = \overline{V}_t + \overline{I}_a r_a + j\overline{I}_d X_d + j\overline{I}_q X_q \tag{1}$$

Here $X_d = x_{al} + X_{md}$ = d-axis synchronous reactance and
$X_q = x_{al} + X_{mq}$ = q-axis synchronous reactance

The two-reaction theory phasor diagram for a salient pole synchronous generator is shown in Fig. 11.2(a). The equivalent circuit diagrams for X_d and X_q are shown in Fig. 11.2 (b) and (c) respectively.

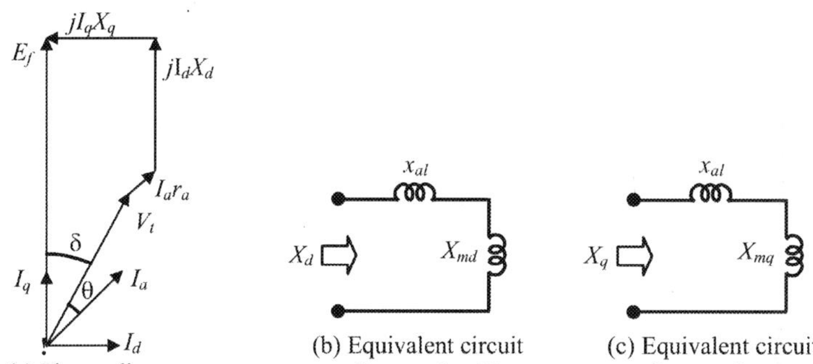

(a) Phasor diagram (b) Equivalent circuit diagram for X_d (c) Equivalent circuit diagram for X_q

Fig. 11.2. Salient pole synchronous generator.

3.2 Classification of Synchronous Machine Reactances

The synchronous machine reactances can be classified into three types, namely,

1. Positive sequence reactance
2. Negative sequence reactance
3. Zero sequence reactance

The synchronous machine has several positive sequence reactances:

1. X_d,
2. X_q,
3. X_d' = d-axis transient reactance
4. X_q' = q-axis transient reactance
5. X_d'' = d-axis sub-transient reactance
6. X_q'' = q-axis sub-transient reactance

The synchronous machine has only one negative sequence reactance, X_2 and only one zero sequence reactance, x_0. The d-axis synchronous reactance X_d is the largest and the zero sequence reactance x_0 is the smallest of all the synchronous machine reactances. The harmonic sequence components other than fundamental component, are obtained from the equation $6n \pm 1$, where $n = 1, 2, 3\ldots$. For example, if $n = 1$, then 7th harmonic gives positive sequence reactance and 5th harmonic gives negative sequence reactance. The zero sequence reactance is obtained from 3rd and triplen harmonic (3^n), i.e., 9th, 27th and so on.

3.3 Positive Sequence Reactances

The magnitude of different positive sequence reactances depend upon:

(a) the angular position of the field structure with respect to the armature mmf wave; and

(b) whether the armature currents are steady or varying.

When both the field and armature mmf wave are almost stationary with respect to each other or when they rotate at the same speed in the same direction, then the reactances associated with the armature are positive sequence reactances.

The armature mmf wave actually consists of a space fundamental component and many other space harmonic components. For a three-phase system, the voltage equations with harmonic can be written as

$$\left.\begin{aligned} V_R &= V_{m1} \sin \omega t + V_{m3} \sin 3\omega t + V_{m5} \sin 5\omega t + \ldots\ldots\ldots \\ V_Y &= V_{m1} \sin(\omega t - 120) + V_{m3} \sin 3(\omega t - 120) + V_{m5} \sin 5(\omega t - 120) + \\ V_B &= V_{m1} \sin(\omega t - 240) + V_{m3} \sin 3(\omega t - 240) + V_{m5} \sin 5(\omega t - 240) + \ldots \end{aligned}\right\} \quad (2)$$

Here V_{m1}, V_{m3}, V_{m5} are the maximum values of the fundamental, third and fifth harmonic components of the applied voltage, respectively.

The fundamental component voltages are:

$$\left.\begin{aligned} V_{R1} &= V_{m1} \sin \omega t \text{ for phase R} \\ V_{Y1} &= V_{m1} \sin(\omega t - 120) \text{ for phase Y} \\ V_{B1} &= V_{m1} \sin(\omega t - 240) \text{ for phase B} \end{aligned}\right\} \quad (3)$$

These voltages form a three-phase balanced system with phase sequence RYB as shown in Fig. 11.3(a).

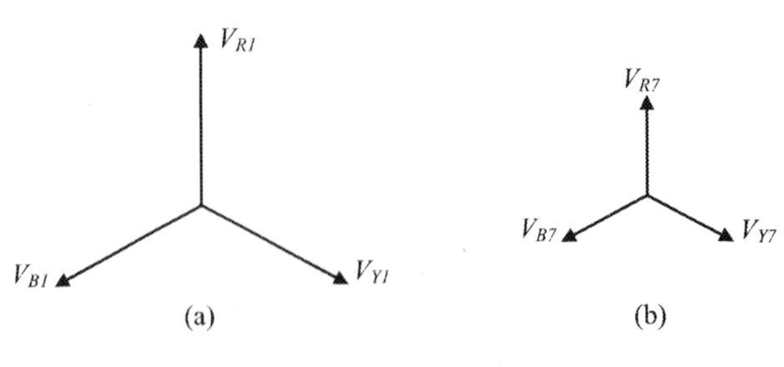

Fig. 11.3. Phasor diagrams for fundamental and seventh harmonic voltages.

The seventh harmonic voltages are:

$$\left.\begin{aligned} V_{R7} &= V_{m7} \sin \omega t \text{ for phase R} \\ V_{Y7} &= V_{m7} \sin(7\omega t - 720) = V_{m7} \sin(\omega t - 120) \text{ for phase Y} \\ V_{B7} &= V_{m7} \sin (7\omega t - 1680) = V_{m7} \sin (\omega t - 240) \text{ for phase B} \end{aligned}\right\} \quad (4)$$

It is seen from the above equations that the seventh harmonic voltages are having time phase displacement of 120° with phase sequence RYB, similar to fundamental components as shown in Fig. 11.3(b). Therefore the seventh harmonic components help in producing the sequence reactances at its synchronous speed similar to fundamental components.

For a low applied voltage and low current, the other harmonic components can be neglected. As far as the harmonic components are considered, the positive sequence reactances are due to:

1. space fundamental component of armature mmf wave
2. seventh harmonic component and so on.

If the field structure is rotated in the same direction and at the same speed.

The reactances of these two windings along with these two axes are discussed below.

3.3.1 Direct-axis Synchronous Reactance, X_d

When the machine is operating under steady state conditions, the magnitude and line of action of resultant armature mmf are fixed with respect to field. This mmf establishes flux, which is shown in Fig. 11.4(a). The armature mmf is $N_a I_a$.

When the peak of steady armature mmf wave is along direct or polar axis, then the situation is as illustrated in Fig. 11.4(b). It is seen from this figure that reluctance, offered to armature flux by two small air gaps along d-axis, is minimum. Under this condition the armature flux linkages per ampere in the armature winding is called **direct axis synchronous inductance L_d**, which is given by:

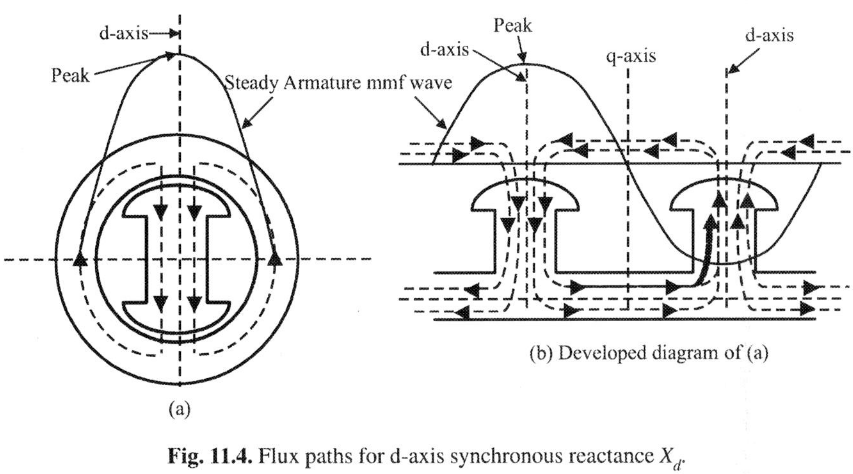

Fig. 11.4. Flux paths for d-axis synchronous reactance X_d.

$$L_d = \frac{\text{Armature flux} \times \text{armature phase turns}}{\text{Armature phase current}} \tag{5}$$

and d-axis synchronous reactance $X_d = \omega L_d = 2\pi f L_d$.

For minimum air gap and minimum reluctance, the armature current, which is required for establishing the **constant** air gap flux, must be MINIMUM (since $\Phi = (N_a I_a / Reluctance)$. The constant applied voltage minus the minimum impedance voltage drop ($I_{aMIN}Z$) gives MAXIMUM armature terminal voltage between the leads and three-phase variac. Thus the d-axis synchronous reactance is given by

$$X_d = \frac{V}{I} = \frac{\text{Maximum armature terminal voltage per phase}}{\text{Minimum armature current per phase}}$$

$$= \frac{V_{aMAX}}{I_{aMIN}}, \text{ if resistance is neglected} \tag{6}$$

Along d-axis the armature mmf is fully spent in establishing d-axis flux only and there is no component on q-axis. Hence, the reactance offered by armature coil is X_d.

Since the relative speed/velocity between armature mmf wave and field structure is zero, the field may be open or closed. But when there is speed difference (or slip) between these two, voltage may be induced across the field winding due to armature flux and field current may affect the steady armature flux when field is closed. Therefore, the ***field is kept open*** during this test.

3.3.2 Quadrature-axis Synchronous Reactance, X_q

When the machine is operating under steady state conditions, the magnitude and line of action of resultant armature mmf are fixed with respect to field. This mmf establishes flux, which is shown in Fig. 11.5(a). The armature mmf is $N_a I_a$.

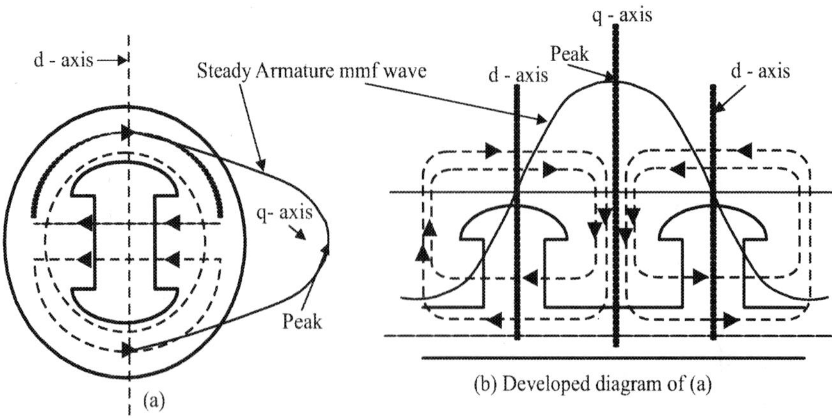

Fig. 11.5. Flux paths for q-axis synchronous reactance X_q.

When the peak of steady armature mmf wave coincides with the quadrature or interpolar axis, then the situation is as illustrated in Fig. 11.5(b). It is seen from this figure that reluctance, offered to armature flux by two large air gaps along q-axis, is maximum. Under this condition the armature flux linkages per ampere in the armature winding is called **quadrature axis synchronous inductance L_q**, which is given by

$$L_q = \frac{\text{Armature flux} \times \text{armature phase turns}}{\text{Armature phase current}} \tag{7}$$

and q-axis synchronous reactance $X_q = \omega L_q = 2\pi f L_q$

For maximum air gap and maximum reluctance, the armature current, which is required for establishing the **constant** air gap flux, must be MAXIMUM (since $\Phi = (N_a I_d / Reluctance)$. The constant applied voltage minus the maximum impedance voltage drop (I_{aMAX} Z) gives MINIMUM armature terminal voltage between the leads and three-phase variac. Thus the q-axis synchronous reactance is given by

$$X_q = \frac{V}{I} = \frac{\text{Minimum armature terminal voltage per phase}}{\text{Maximum armature current per phase}}$$

$$= \frac{V_{aMIN}}{I_{aMAX}}, \text{ if resistance is neglected} \tag{8}$$

The reactance offered by armature coil along q-axis is X_q because the armature mmf is fully spent in establishing q-axis flux only and there is no component on d-axis. The voltage and current are changing from min to max values. If analog voltmeter and ammeter are connected, then pointers of these meters swing between min *t* max values. The ammeter pointer swing is wide and and the voltmeter has only small swing because of low impedance voltage drop.

3.3.3 Armature Leakage Reactance, x_a

The total flux produced by the armature current consists of two components – one component crosses the air gap whereas the second component does not cross the air gap and links armature only. The leakage flux may be due to slot leakage, tooth head leakage and coil end or overhang leakage. The component of the total flux that links armature alone is called the **armature leakage flux**, which gives rise to the **armature leakage reactance x_a**. All the positive sequence reactances have x_a as one part of it. For example, $X_d = x_a + X_{md}$ and $X_q = x_a + X_{mq}$. Here X_{md} and X_{mq} are the armature magnetizing reactances or mutual reactances along d and q-axes respectively. For convenience, armature leakage reactance x_a is considered equal for both direct and quadrature axes reactances.

3.4 Measurement Methods

The reactances x_d and x_q can be measured by the following methods.

3.4.1 Measurement of X_d from OCC and SCC

Refer to phasor diagram shown in Fig. 11.2 (a). The excitation voltage or open circuit voltage E_f is given by

$$E_f = 4.44 f \Phi_f N_{ph} K_w = K \omega I_f \tag{9}$$

By neglecting the resistance, r_a, we can write

$$V_t \sin \delta = I_q X_q \tag{10}$$

Under short circuit condition, the terminal voltage, V_t is zero.

$$0 = I_q X_q \text{ or } I_q = 0 \tag{11}$$

Now the short circuit current is

$$I_{sc} = I_d + jI_q = I_d \tag{12}$$

Thus under short circuit condition, V_t is zero and when r_a is neglected, the short circuit armature current I_{sc} is equal to I_d, which is along d-axis. Therefore, d-axis parameters, X_d, X_d' and X_d'' are involved in short circuit analysis and short circuit current is dependent on them. The modified phasor diagram under short circuit condition is shown in Fig. 11.6.

From the phasor diagram it is clear that

$$E_f = I_d X_d \text{ or } I_d = I_{sc} \propto E_f \propto I_f \tag{13}$$

$$X_d = \frac{E_f}{I_d} = \left|\frac{E_f}{I_{sc}}\right| \text{ at same } I_f \tag{14}$$

$$X_d = \frac{\text{Open circuit voltage for a given field current}}{\text{Short circuit current for the same field current}} \tag{15}$$

The open circuit voltage E_f is obtained from OCC and the short circuit current I_{sc} is obtained from SCC, for a given field current.

$$I_{sc} = \frac{E_f}{X_d} = \frac{K_w I_f}{wLd} = \frac{KI_f}{L_d} = K_1 I_f \tag{16}$$

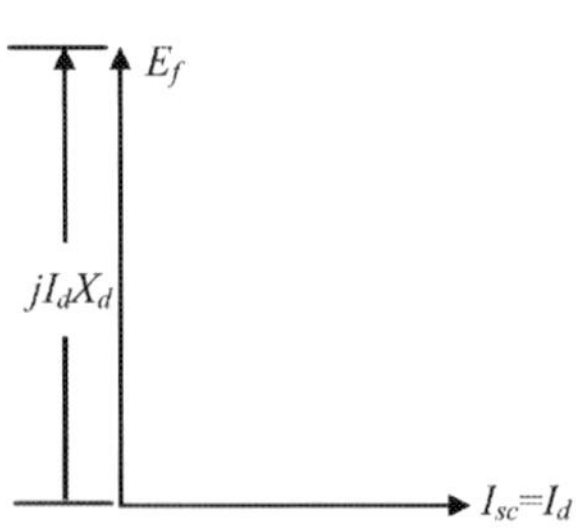

Fig. 11.6. Phasor diagram for salient pole alternator under short circuit condition with $r_a = 0$.

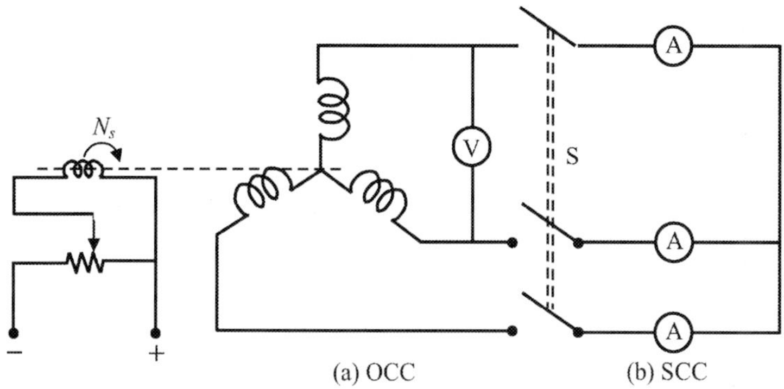

Fig. 11.7. Schematic diagram for open circuit and short circuit test.

Thus the short circuit current is directly proportional to the field current, I_f because L_d is synchronous machine constant. For short circuit test it is not necessary that the alternator speed should be equal to the synchronous speed. But at lower speed the reactance decreases and becomes less than the resistance so that the short circuit current will not vary linearly with the field current.

Under the short circuit condition, since V_t is zero, the air gap voltage is sufficient to overcome the leakage impedance drop I_{se} $(r_a + jx_a l)$. Since $x_a l$ is more than r_a the short circuit current logs the air gap voltage by nearly 90°. The amature reaction mmf is almost opposite to field mmf and demagnetizing in nature. The resultant mmf, resultant air gap flux are reduced and there is no saturation under short circuit condition. Therefore the short circuit characteristic is a straight line passing through the ogive, as shown in Fig. 11.8.

Since the leakage reactance is small, a low voltage, generally 4 to 8% of the rated voltage is required to circulate the rated current under short circuit condition. Due to low voltage applied, the machine is under unsaturated region and as a result SCC is a straight line.

The open circuit characteristic (OCC) is obtained by running the alternator at synchronous speed and measuring the open circuit voltage, E_f, across the armature

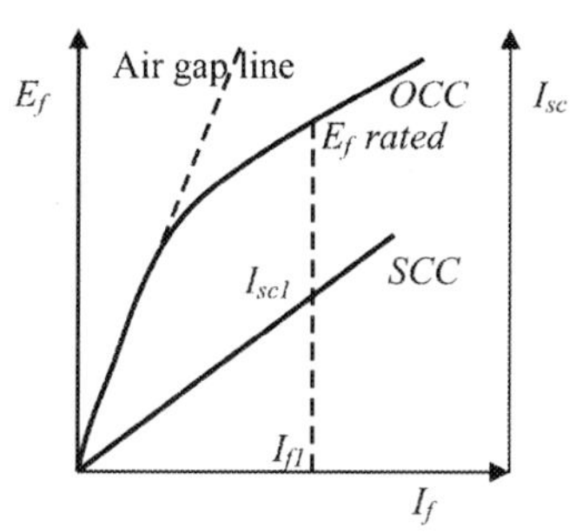

Fig. 11.8. Open circuit and short circuit characteristics.

by varying the field current, I_f. Thus, OCC is a graph between open circuit voltage and field current. The circuit diagram for obtaining OCC and SCC is shown in Fig. 11.7. The open circuit test is performed with switch S open. With the variation of field current, the armature voltage is noted. The OCC is shown in Fig. 11.8, which is a magnetization curve.

At low value of I_f, when iron is in unsaturated state, the OCC is almost linear and mmf applied is mainly consumed in establishing air gap flux, the reluctance of the iron path is almost negligible. The straight line part of the OCC, if extended, is called ***air gap line***, as shown in Fig. 11.8. The open circuit test conducted at rated voltage gives the constant loss of the synchronous machine comprising core loss and mechanical loss due to friction and windage.

The short circuit test is performed by running the machine at synchronous speed and closing the switch *S* with field current zero. The field excitation is gradually increased till the armature current is equal to rated value. The SCC is a straight line as per equation 10 and is shown in Fig. 11.8. Since under short circuit condition the machine is highly under-excited, the losses comprise mechanical loss and copper loss in the resistance of the armature, the iron loss being negligible. The value of X_d as per the equation 12, is given by

$$X_d = \left| \frac{E_{f\,rated}}{I_{sc1}} \right|_{I_{f1}} \tag{17}$$

In unsaturated region, the value of X_d remains constant then decreases in the saturated region.

X_d can also be calculated from symmetrical short circuit armature current waveform. This plot can be obtained from the oscillogram of the short circuit current in any one of the phases with zero DC off-set, as shown in Fig. 11.9. The current is decaying due to circuit time constant. The short circuit current is along d-axis as per Fig. 11.6, therefore short circuit current equation consists of the d-axis parameters X_d, X_d', X_d''. The three periods, i.e., sub-transient, transient and steady state are indicated on the envelop. The decaying of sub-transient current is

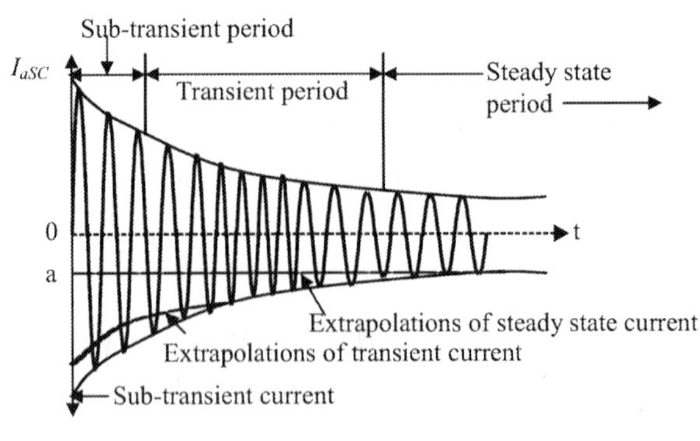

Fig. 11.9. Symmetrical waveform for armature short circuit current of an alternator.

governed by the damper winding time constant, T_d''(d-axis short circuit sub-transient time constant). The decaying of transient current is governed by field winding time constant, T_d' (d-axis short circuit transient time constant). Under steady state condition, the current is dependent on X_d. Therefore, X_d can be calculated from the equation

$$X_d = \frac{E_{f\,rated}}{0a/\sqrt{2}} \tag{18}$$

The short circuit current waveform is obtained by connecting CRO across the rheostat, which is connected in series with any one phase.

3.4.2 Measurement of X_q from Maximum-Lagging-Current Test

The equation of power per phase for salient pole synchronous machine is given by (See Chapter 10, equation 13)

$$P_c = P_m = P = \frac{E_f V_t \sin\delta}{X_d} + \frac{V_t^2}{2}\left(\frac{1}{X_q} - \frac{1}{X_d}\right)\sin 2\delta \tag{19}$$

If the excitation is made negative, then the power is given by

$$P = \underbrace{\frac{-E_f V_t \sin\delta}{X_d}}_{I} + \underbrace{\frac{V_t^2}{2}\left(\frac{1}{X_q} - \frac{1}{X_d}\right)}_{II}\sin 2\delta \tag{20}$$

The electromagnetic power (I) and reluctance power (II) are opposing each other. The magnitude of negative excitation is increased till steady state limit is reached when the motor is just on the verge of losing synchronism. The maximum negative excitation is obtained by equating P to zero.

$$\frac{E_f V_t \sin\delta}{X_d} = \frac{V_t^2}{2}\left(\frac{1}{X_q} - \frac{1}{X_d}\right) 2\sin\delta\cos\delta$$

$$E_f = V_t\left(\frac{X_d - X_q}{X_q}\right)\cos\delta \tag{21}$$

Since $P = 0$, $\delta = 0$, therefore maximum negative excitation voltage is

$$E_{f\,MIN} = V_t\left(\frac{X_d - X_q}{X_q}\right) \tag{22}$$

The reactive power is given by (See Chapter 10, equation 15)

$$Q = \frac{V_t^2}{X_d} + V_t^2\left(\frac{1}{X_q} - \frac{1}{X_d}\right)\sin^2\delta - \frac{E_f V_t}{X_d}\cos\delta \tag{23}$$

For negative excitation

$$Q = \frac{V_t^2}{X_d} + V_t^2\left(\frac{1}{X_q} - \frac{1}{X_d}\right)\sin^2\delta + \frac{E_f V_t}{X_d}\cos\delta \tag{24}$$

For maximum negative excitation $E_{f.MIN}$, P = 0, δ = 0 therefore,

$$Q = \frac{V_t^2}{X_d} + \frac{E_{f\,MIN} V_t}{X_d}$$

$$= \frac{V_t^2}{X_d} + V_t \frac{X_d - X_q}{X_q} \times \frac{V_t}{X_d}$$

$$= \frac{V_t^2}{X_q} \tag{25}$$

Since Q is positive, motor absorbs reactive power from the power system. For zero power, $V_t I_a \cos\theta = 0$, V_t and I_a can't be zero, so cos θ = 0 and θ = 90°. With this the reactive power is

$$Q = V_t I_a \sin 90 = V_t I_a = \frac{V_t^2}{X_q}$$

$$\therefore\ X_q = \frac{V_t}{I_a} \tag{26}$$

Thus, as per equation 26, X_q can be obtained since the current I_a is lagging to the voltage V_t by 90° this test is called maximum lagging current test.

The schematic diagram for maximum lagging current test is shown in Fig.11.10. In this test the salient pole synchronous motor is run on no load, with rated voltage

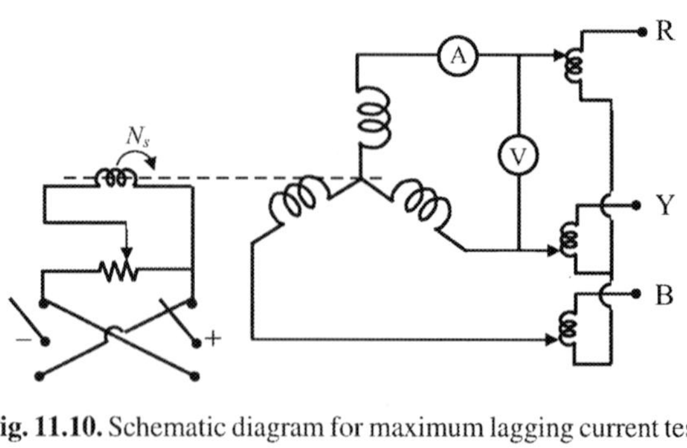

Fig. 11.10. Schematic diagram for maximum lagging current test.

across the armature. The field current is gradually reduced to zero, so $E_f = 0$ and the motor runs as a reluctance motor at synchronous speed. Now the polarity of E_f is reversed and then increased. At some value of negative excitation, rotor falls out of step momentarily and armature current increases. The armature terminal voltage and maximum stable current are noted and as per equation 25, X_q is calculated.

3.4.3 Measurement of X_q from Pull-out test

The unsaturated value of qudrature axis synchronous reactance X_q can be determined by this method. The synchronous machine is started as a synchronous motor with rated voltage applied to the armature. The field excitation is gradually reduced to zero. The input power at this stage is reluctance power which is given by

$$P = \frac{r_L^2}{z}\left(\frac{1}{x_q} - \frac{1}{x_d}\right)\sin 2\delta \qquad (27)$$

Now slowly decrease the armature voltage untill the motor drops out of synchronism. This can be viewed by seeing the deflection in the ammeter and wattmeter. At this instant the reluctance power is maximum possible for the corresponding impressed reduced voltage. Note the power and reduced voltage. By knowing X_d for *OCC* and *SCC* as per equation (27) λ_q can be calculated.

3.4.4 Measurement of X_d and X_q from Slip Test

The voltage and current are changing from minimum to maximum values as per equations 6 and 8. If analog voltmeter and ammeter are connected, then pointers of these meters swing between minimum and maximum values. The ammeter pointer swing is wide and the voltmeter has only small swing because of the low impedance voltage drop.

The steady armature mmf wave rotates at synchronous speed. If rotor or field structure is rotated by prime mover or motor, at a speed other than synchronous speed, the relative speed between these two speeds is ***slip speed***, and the test where this difference of speed is maintained, is called ***slip test***. For the measurement of X_d and X_q, which are positive sequence reactances, the armature mmf wave and field structure must rotate in the same direction. If the field poles revolve in a direction opposite to the rotating mmf wave, negative sequence reactance would be measured. The schematic diagram for the slip test is shown in Fig. 11.11.

While performing this test, slip should be made as small as possible. With small slip, there is more time (t = 1/sf) available for swinging the pointers of meters. So readings can be taken easily with minimum error and better results can be obtained.

The reduced voltage (around 25% of rated value) is applied to the armature to have only fundamental component of voltage with no harmonics for producing

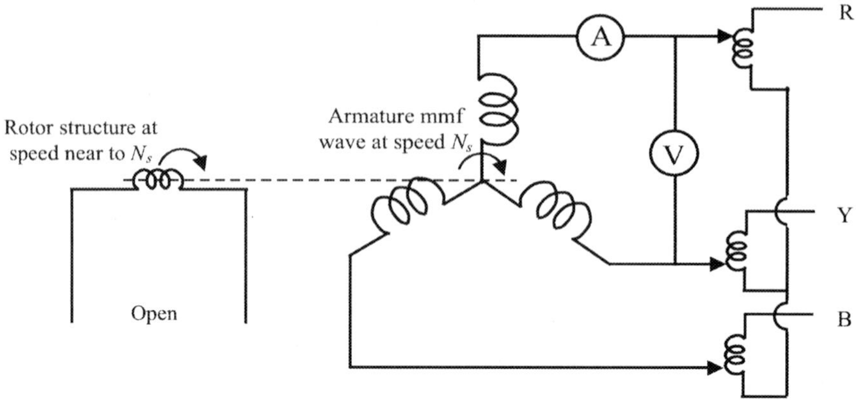

Fig. 11.11. Connection diagram for slip test for the measurement of X_d and X_q.

steady armature mmf. Again armature current is low with low voltage and small slip. With low current, reluctance torque (due to the presence of damper winding) is low in salient pole machine and synchronization is avoided. With higher armature current, reluctance torque is more and there may be synchronization, i.e., equal speeds of armature mmf and field poles.

With equal speeds, peak of armature mmf wave may be either along d-axis or along q-axis. There will not be any swinging of the pointers of the meters and readings cannot be taken.

With reduced voltage, X_d and X_q are obtained under unsaturated region and not under saturated region of BH curve but the error is cancelled out if ratio is taken. Therefore, slip test is suitable for obtaining the ratio X_q/X_d.

Due to the difference in speeds, at one instant rotating armature mmf wave is along d-axis and at other instant that is along q-axis. The time interval between these two situations is that required to turn the rotor through 90° electrical at slip speed. The analog voltmeter and ammeter are noted and as per equations 6 and 8, X_d and X_q are calculated.

The air gap along d-axis is less than the air gap along q-axis, therefore it can be concluded that $X_d > X_q$ for salient pole synchronous machine. Normally the ratio is between 0.6 and 0.7.

Since cylindrical rotor machines have uniform air gap, $X_d = X_q = X_a$ where X_a is synchronous reactance. Strictly speaking X_q is slightly less than X_d because of the presence of air gap along q-axis due to the rotor slots and its opening. The ratio of X_q/X_d is not equal to 1.0 but 0.95 to 0.98.

In this test the field circuit is kept open. If field is shorted, the slip frequency current will flow through it and it will disturb the steady state armature mmf wave. So we may not get the readings for X_d and X_q. If voltmeter is connected across the open field terminals, then voltmeter will read slip frequency voltage as shown in Fig. 11.12.

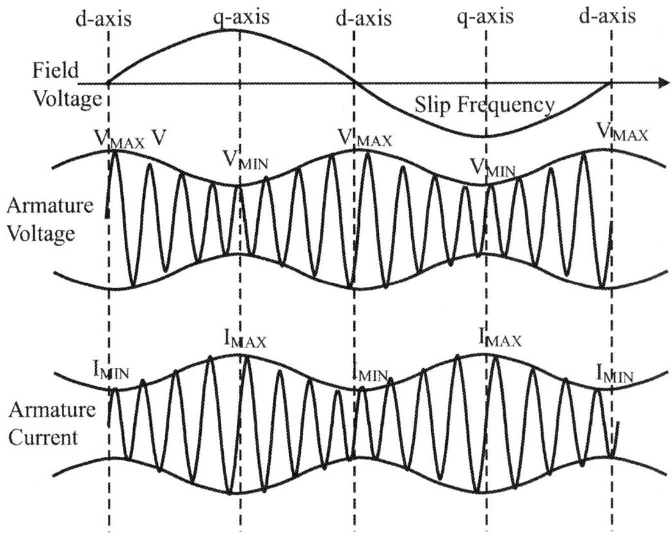

Fig. 11.12. Voltage and current waveform in slip test.

Along d-axis, the armature flux passing through open field winding is maximum, therefore induced field emf is zero. Along q-axis, the armature flux passing through open field winding is zero; therefore induced field emf is maximum. As the flux set up by armature current moves fast the rotor field, the flux linkage of the field vary and an emf of twice the slip frequency is induced in it. The voltmeter and ammeter oscillate at twice the slip frequency.

4. Apparatus Required

1. Synchronous machine: rating: (For example, 7.5 kVA, 440 V, 1500 rpm)
2. DC shunt motor (prime mover): rating: 5 kW, 230 V, 1600 rpm
3. Three-phase dimmerstat: rating: 7.5 kVA, 440 V
4. AC Ammeter: one, 0-10 amps.
5. AC Voltmeter: one, 0-75 volts.
6. Rheostat: one, 300 ohms, 2 amps.
7. Tachometer or speedometer: one

5. CIRCUIT DIAGRAM

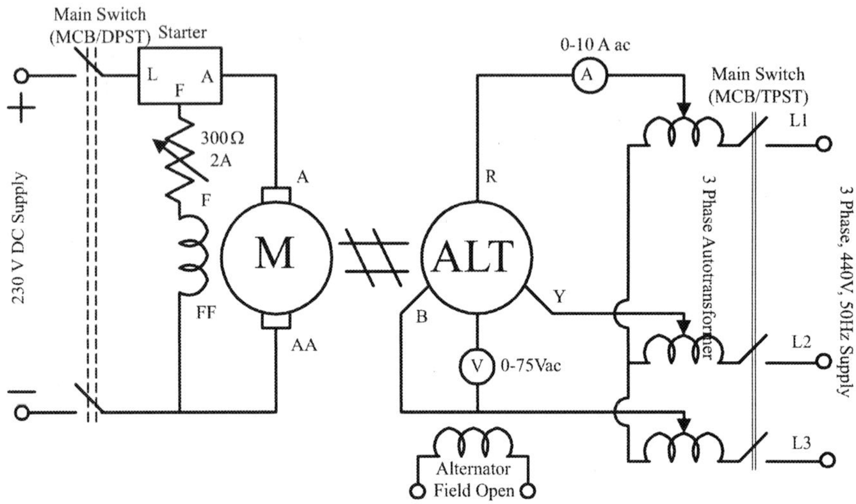

Fig. 11.13. Experimental set-up for determining the ratio X_q/X_d by slip-test.

6. PROCEDURE

1. Make the connections as shown in Fig. 11.13. Set the variac output to zero.
2. If the synchronous machine is not having damper winding, housed in the pole shoe, short the field winding to produce torque. Check the direction of rotation of rotor by increasing the armature voltage to suitable value by the use of variac.
3. Decrease the armature voltage to zero, disconnect the three-phase supply and open the field winding.
4. Set up the speed control rheostat of DC motor, to minimum speed point.
5. Switch on the DC supply with the help of main switch or MCB. Move the handle of starter slowly and start the DC motor. The direction of rotation should be same as that of arm field.
6. Adjust the speed of DC motor near to synchronous speed of alternator but less than synchronous speed (1480 rpm if synchronous speed is 1500 rpm).
7. Apply reduced voltage to the armature by variac.
8. Observe the swings or variations in voltmeter and ammeter readings.
9. Adjust the speed of the DC motor FINELY to get maximum swings in ammeter and voltmeter pointers.
10. Note minimum and maximum readings of voltage and current.

11. Take additional sets of readings by adjusting different variac outputs and at different speeds.
12. Now adjust the DC motor speed to a value little higher than synchronous speed (1510 rpm) and take similar readings as above.

7. OBSERVATIONS AND CALCULATION

Sr. No.	Speed	Armature Voltage		Amature Current		$X_d = \frac{V_{aMAX}}{I_{aMIN}}$	$X_q = \frac{V_{aMIN}}{I_{aMAX}}$	$\frac{X_q}{X_d}$
		Minimum $V_{a\,MIN}$	Maximum V_{aMAX}	Minimum $I_{a\,MIN}$	Maximum $I_{a\,MAX}$			
1.	1480							
2.	1490							
3.	1510							
4.	1520							

8. RESULT

For salient pole synchronous machine, the ratio of X_q/X_d is found to be...............

9. CONCLUSION

The value of X_d is more than X_q.

10. POST-EXPERIMENTAL QUIZ

1. For salient pole synchronous machine, why X_d is more than X_q?
2. For cylindrical synchronous machine why the ratio of X_q/X_d is not equal to 1?
3. Why field winding is kept open?
4. Why reduced voltage is given to the armature?

CHAPTER 12

To Study the Synchronization of an Alternator with an Infinite Bus by "Dark Lamp Method"

1. OBJECTIVE

The main objectives of this experiment are:

1. to study the process of synchronization.
2. to study the sharing of load by alternator after synchronization.

2. PRE-EXPERIMENTAL QUIZ

1. What do you mean by synchronization?
2. Explain infinite bus.
3. List the conditions for synchronization?

3. THEORY

Actually, electricity supply of a state or a country is fed by large number of alternators operating in parallel. The nuclear, thermal, hydro, gas and unconventional power plants are generating power stations. All these stations are far away from each other and are interconnected by long transmission lines and transformers to a power grid, which is called the national grid. The national level grid consists of hundreds of generators and in the length thousands of kilometre of transmission lines. There are a number of good reasons for this practice of parallel operation. The demand of electrical energy is continuously growing due to electrical utilities and industries, therefore it is necessary to increase the generating power stations and then connect them to the power grid.

3.1 Regional Electricity Boards

The Government of India had formed seven regional electricity boards in the country to supply electrical power to the customer without interruption. These regional electricity boards are as follows:

1. Eastern Regional Electricity Board (EREB):
   EREB consists of Bihar, Jharkhand, West Bengal, Orissa and Sikkim.
2. Western Regional Electricity Board (WREB):
   WREB consists of Maharashtra, Madhya Pradesh, Gujarat, Chhattisgarh, Goa, Daman-Diu and Nagar Haveli.
3. Northern Regional Electricity Board (NREB):
   NREB consists of Jammu & Kashmir, Himachal Pradesh, Punjab, Haryana, Chandigarh, Delhi, Uttarakhand, Uttar Pradesh and Rajasthan.
4. Southern Regional Electricity Board (SREB):
   SREB consists of Andhra Pradesh, Karnataka, Kerala, Tamil Nadu and Pondicherry.
5. North-East Regional Electricity Board (NEREB):
   NEREB consists of Assam, Arunachal Pradesh, Nagaland, Meghalaya, Manipur, Mizoram and Tripura.
6. Andaman and Nicobar Electricity Board
7. Lakshadweep Electricity Board.

3.2 Infinite Bus

A bus bar (national grid) which maintains constant voltage and constant frequency irrespective of the load variations on it, is called an *infinite bus*. A large number of alternators interconnected together to form a supply system, may be regarded as an infinite bus. An infinite bus behaves like a voltage source having ***zero internal impedance*** and ***infinite rotational inertia***. The zero internal impedance maintains the voltage constant and infinite inertia maintains speed and frequency constant under load variations. Therefore, an alternator switched on to or off, the infinite bus does not cause any change in the voltage and frequency of the system.

3.3 Synchronization

Synchronizing is the process of **switching on** of an alternator to another alternator or bus bar, **or connecting** two alternators in parallel, without any interruption. The synchronous machine which is to be synchronized is called an *incoming machine*. There are certain **conditions** to be satisfied before a generator can be connected in parallel to other generators or to the infinite bus.

1. The terminal **voltage** of the incoming machine and the already running alternator or bus-bars or infinite bus must be **equal**.
2. The **frequency** of the two voltage sources must be **equal,** i.e., machine speed should be close to synchronous.
3. The **phase sequence** must be same and there must be **same phase to phase** connection.

The first condition of voltage of the incoming machine is made equal to the infinite bus by field excitation. The equality can be checked by a single voltmeter connecting first to the infinite bus and then to the incoming generator. The second condition of frequency of the incoming machine is made equal to the infinite bus by prime mover. The conditions of frequency and phase sequence are checked with the help of synchronizing lamps. These lamps are connected across the ***triple pole double throw*** switch which is called synchronizing switch in case of three-phase alternator.

Departure from the above conditions causes the appearance of current and power surges in the system, which are accompanied by undesirable electro-mechanical oscillations of the rotor.

Practically, at the time of synchronization, it is desirable that the voltage and frequency of the incoming machine should be slightly more than the infinite bus. At the time of synchronization the incoming machine is suddenly loaded, due to which the voltage and frequency of the incoming machine decreases and machine may operate as a motor. So instead of sharing the load of infinite bus by operating as a generator, machine acts as a load on infinite bus by operating as a motor. This is not a desirable operation. This is explained below by drawing phasor diagram.

Consider an infinite bus, with voltage E_f, supplying a current I_a. The generator with voltage E_{f1} is synchronized with infinite bus with speed slightly less than infinite bus. The phasor diagram for alternators with internal circuit is shown in Fig. 12.1(a). At the time of synchronizing, E_{f1} falls back, the voltage E_c appears in the local circuit between two alternators. The circulating current I_c flows which lags the voltage E_c by 90° due to inductive circuit. The resultant current of the infinite bus is I_a' which is more than I_a and the power of infinite bus is $E_f I_a'$ cos (angle between E_f & I_a') which is more than the earlier power. This shows that already loaded generator gets loaded further. The power of incoming generator E_{f1} I_c cos (angle between E_{f1} & I_c) is negative. This means the incoming generator operates as a motor which is an undesirable operation.

Now consider the incoming generator is synchronized with a speed slightly more than infinite bus. E_{f1} gets ahead with respect to earlier position. The resultant current of the infinite bus is I_a' which is less than I_a and the power of infinite bus is $E_f I_a'$ cos (angle between E_f & I_a') which is also less than the earlier power. This

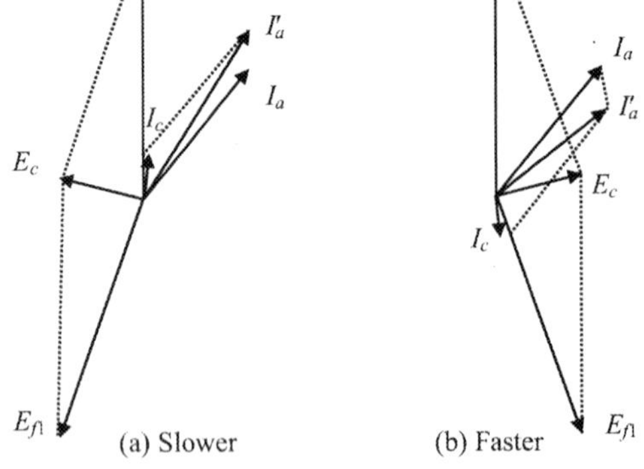

Fig. 12.1. Phasor diagram of alternators at the time of synchronization.

shows that already loaded generator gets relieved. The power of incoming generator $E_{f1} I_c$ cos (angle between E_{f1} & I_c) is positive. This means the incoming generator operates as a generator and shares some load which is a desirable operation.

3.4 Reasons for Synchronization

The ***reasons*** for interconnecting the alternators in parallel are given below:

3.4.1 Efficiency and Operating Cost

With several units in parallel, the number of units operating at a time can be disconnected or connected depending upon the electricity demand. In this manner, it is possible to operate the alternators near their full load with maximum efficiency, therefore reduced operating cost. A single unit will have to operate at a reduced load for certain periods of the day; consequently it has lower operating efficiency and therefore, increased operating costs.

Consider a single generator of 1000 MW and five generators, each of 200 MW connected in parallel. If there is load of 800 MW, then efficiency of a single generator is 80%. The efficiency of five generators set is 100% with the disconnection of any one generator.

3.4.2 Maintenance and Repair

In order to carry out regular and routine maintenance and inspection at the power stations, a unit must be shut down for a certain period. It is possible only when there are several small units connected in parallel. It is more convenient and economical also because the entire load of the system can be supplied by remaining generators at 10 to 20% more than their capacity for same time. The cost of standby unit is also less when several units are installed.

If there is only one generator, there would not be electrical supply to the load during the maintenance and inspection time.

3.4.3 Economy and Utility

The cost of generation is higher for nuclear and lower for hydro-generating stations. The interconnection of all stations enables optimum economy and utility depending on the shut down of the stations when electricity demand is less. Generally, hydro-stations are run for base load, thermal stations are run for moderate load and when there is heavy load or peak load nuclear stations are turned on. But nowaday, the total load is more than total power generated, that's why all the generators are kept on.

3.4.4 Continuity of Services

In case of a fault on any unit, it can be shut down for repairs and power supply can still be maintained continuous to the consumer with the operation of other healthy stations.

3.5 Methods of Synchronization

There are three methods of synchronizing the alternators.

1. Dark lamp method, 2. Bright lamp method, 3. Synchroscope method.

3.5.1 Dark Lamp Method

This is the simplest method of synchronizing an incoming machine with another alternator or bus bars. For simplicity first consider the synchronization of single phase alternator with single phase bus bar or in parallel with another alternator. Fig. 12.2. illustrates the connection scheme, where the lamps are connected across the synchronizing switch S_1 (double pole double throw).The incoming machine is run at synchronous speed and field current is adjusted in such a way that its terminal voltage becomes equal to single phase bus bar voltage. The frequency of an incoming machine is made equal to single phase bus bar by adjusting the speed of prime mover. If all the conditions of synchronization are satisfied, then all the lamps are dark, that's why this method is called ***dark lamp method***.

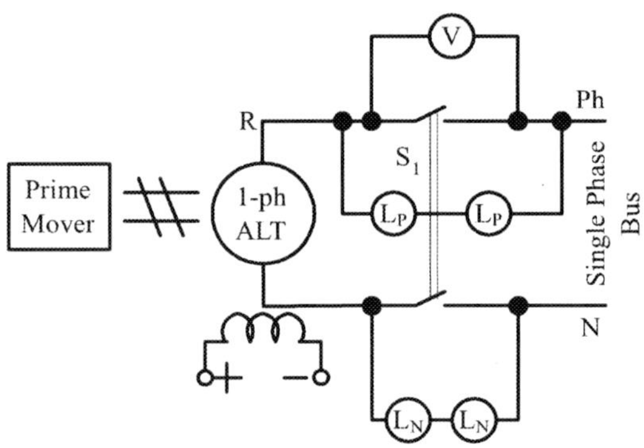

Fig. 12.2. Synchronizing by dark lamp method for single phase alternator.

If the frequencies of the alternators are not equal, the phase angle between their voltages will be continuously changing and therefore the current through the synchronizing lamps and through the local circuit will be changing, resulting in flickering of lamps. The frequency of flickering of lamps will be equal to the difference of frequencies of the two alternators. For greater the difference in frequencies, greater will be the frequency of flickering and for lesser the difference in frequencies, lesser will be the frequency of flickering. In the middle of the dark period the two voltages will be in phase opposition in local circuit and the synchronizing switch S_1 is turned on.

Now consider the synchronization of three-phase alternator with three-phase bus bar or in parallel with another alternator Fig. 12.3. illustrates the connection scheme, where the two lamps are connected across the synchronizing switch S_1 (triple pole double throw) for each phase.

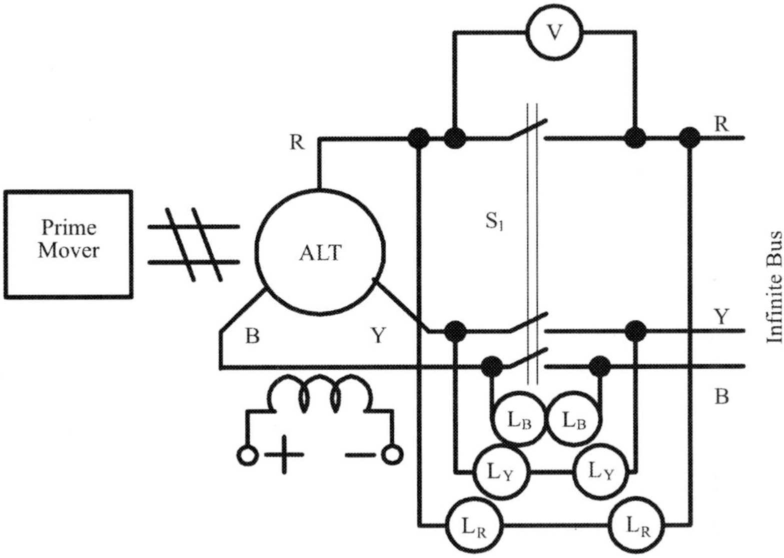

Fig. 12.3. Synchronizing by the dark lamp method for three-phase alternator.

Similar to single-phase alternator, the voltage and frequency of incoming alternator is made equal to infinite bus and when the phase-to-phase connection is correct, the lamps are dark. In the middle of the dark period, the synchronizing switch S_1 is closed. Each time the voltages of R, Y, B of alternator and infinite bus **coincide** and the lamps are **DARK**. When the voltages are in opposite, then the lamps are brightest, and the voltage across each lamp is $2 \times V_{ph}$. Therefore, two lamps of 230 volts are connected in series.

Suppose the frequencies of the infinite bus and alternator are f and f_1 respectively. Then for equal voltages, the voltage across the synchronizing switch S_1 is

$$V_{S1} = V_m \sin \omega t - V_m \sin \omega_1 t$$

$$= 2V_m \left[\cos\left\{ \frac{f_1 + f}{2}(2\pi t) \right\} \times \sin\left\{ \frac{f_1 - f}{2}(2\pi t) \right\} \right] \tag{1}$$

The low frequency $\frac{f_1 - f}{2}$ is known as the ***beat frequency*** and the high frequency $\frac{f_1 + f}{2}$ is known as the ***carrier frequency.*** The voltage waveforms with beat and carrier frequencies are shown in Fig. 12.4.

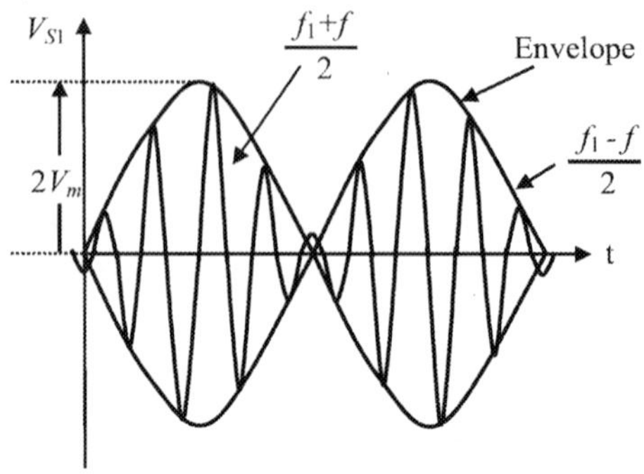

Fig. 12.4. Voltage waveforms with beat and carrier frequencies across switch S_1.

The voltage impressed on the lamps has a frequency equal to the carrier frequency $\frac{f_1+f}{2}$ and the equation for the envelope is $2V_m \sin\frac{f_1-f}{2}(2\pi t)$. Since the lamps are bright once in each half of the voltage wave, the lamps flicker (f_1-f) times in one second.

If the three lamps do not flicker in unison, the phase sequence will be **improper**. This can be rectified by interchanging any two leads either of incoming machine or of infinite bus.

The main ***advantages*** of this method are:

1. It is simple
2. It is quite inexpensive
3. No phase–sequence indicator is required

The ***disadvantages*** are:

1. The lamps become dark at about one-third of their rated voltage. In view of this, there is a possibility of closing the switch S_1 even when there is a considerable voltage difference. This voltage across the switch S_1 will give rise to flow of large current and power, due to which there may be undesirable rotor oscillations.
2. This method does not indicate whether the incoming machine is running too fast or too slow. However, this can be rectified. With the increase of speed of the incoming machine, if flickering increases, then it means that incoming machine is already running faster and this is desirable. For synchronizing, decrease the speed of incoming machine till the lamps flicker very slowly and then close the switch S_1.

With the increase of speed of the incoming machine, if flickering decreases, then it means that incoming machine is already running slower and this is not desirable. For synchronizing, increase the speed of incoming machine till there is no lamp flicker or flicker very slowly and then close the switch S_1.

The above disadvantages are eliminated in the bright lamp method.

3.5.2 Bright Lamp Method (Siemens and Halske)

Figure 12.5 shows the connection scheme for synchronization of single-phase alternator by the bright lamp method. The lamps connections are crossing over as compared to the dark lamp method. If all the conditions of synchronization are satisfied, then all the lamps are bright, that's why this method is called ***bright lamp method***. Under this condition, the synchronizing switch S_1 is closed.

Figure 12.6 illustrates the connection scheme for three-phase alternator. The terminals of two lamps on incoming machine side are interchanged. If all conditions of synchronization are satisfied, then ***two lamps***, of whose terminals are interchanged, are ***BRIGHT*** and third ***lamp is dark***. Under this condition, the switch S_1 is closed. The disadvantages of dark lamp method are removed.

Figure 12.7 shows the phasor diagram of the voltages of infinite bus and generator. The infinite bus voltages are V_{RI}, V_{YI} and V_{BI} as shown by thick lines, and the generator voltages are V_{Rg}, V_{Yg} and V_{Bg} as shown by dotted lines. The names of lamps L_R, L_Y and L_B are given as per infinite bus since the infinite bus voltage and frequency are remaining constant.

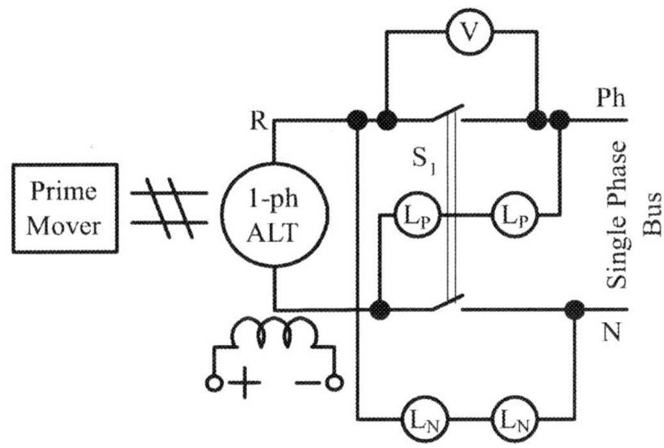

Fig.12.5. Synchronizing by bright lamp method for single phase alternator.

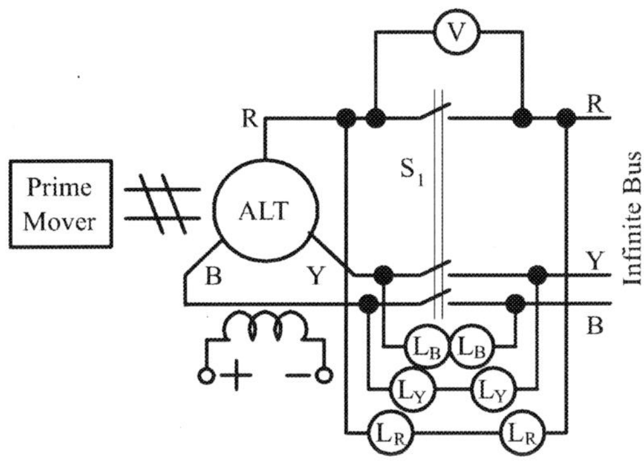

Fig. 12.6. Synchronizing by bright lamp method for three-phase alternator.

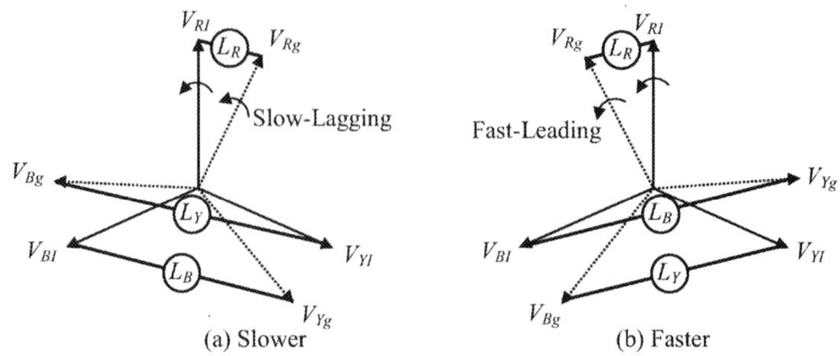

Fig. 12.7. Phasor diagram for bright lamp method.

If the generator is slower than the infinite bus, then the voltage phasors of generator lag to the infinite voltage phasors, as shown in Fig. 12.7(a). First consider the increase in voltage of lamp L_R, so that lamp glows. Then the voltage of lamp L_Y increases and glows and in sequence lamp L_B glows. Therefore, for slower generator the flickering sequence of the lamps is *RYB*.

If the generator is faster than the infinite bus, then the voltage phasors of generator lead to the infinite voltage phasors, as shown in Fig. 12.7 (b). First consider the increase in voltage of lamp L_R, so that lamp glows. Then the voltage of lamp L_B increases and glows and in sequence lamp L_Y glows. Therefore, for faster generator the flickering sequence of the lamps is *RBY*.

As per equation 1, the speed of glowing of lamps is equal to the difference between the two frequencies.

3.5.3 Synchroscope (Loncoln)

Synchroscope is an instrument, fitted with a rotating pointer to indicate the correct instant of closing the switch S_1. For rotating pointer a small motor, with field winding and armature winding is used. The field winding terminals (*B*1, *B*2) are connected to the infinite bus between two lines. The armature has two windings arranged in space quadrature, with terminals *M*1, *M*2 and are connected to incoming generator between the same two lines as that of infinite bus. An elementary connection scheme by synchroscope is shown in Fig. 12.9.

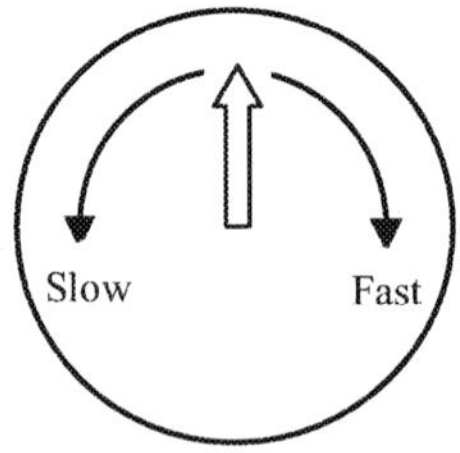

Fig. 12.8. Synchroscope.

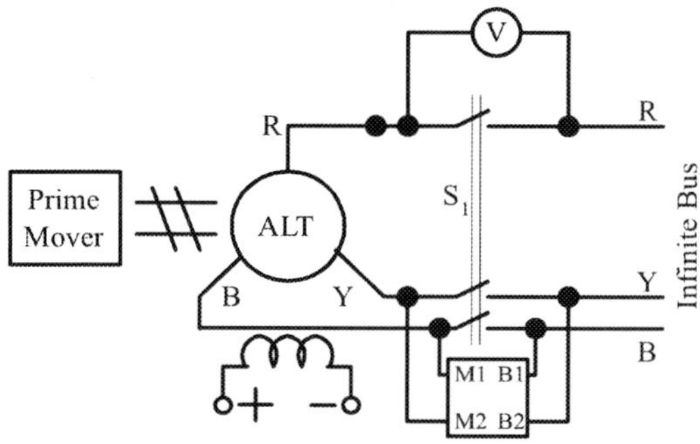

Fig. 12.9. Synchronizing by a synchroscope.

Depending on the phase angle between the infinite bus and generator voltages, torque is developed. If the generator voltage leads, torque is positive and pointer rotates towards fast side (clockwise). This indicates incoming machine is running faster which is a desirable condition for synchronization. If the generator voltage lags, torque is negative and pointer rotates towards slow side (anticlockwise). This indicates incoming machine is running slower which is not a desirable condition for synchronization. The pointer rotates at a speed proportional to the difference in the two frequencies. When the rotating pointer, with very slow speed, reaches the vertical position, in fast direction, then the switch must be closed. If the difference in frequencies or speeds is large, then the pointer does not rotate but oscillates.

Since synchroscope is a single-phase instrument, phase-sequence indicator is required to check phase sequence.

The best synchronization is carried out if both the lamps and synchroscope are used together. Automatic synchronization devices are also available which perform the entire process automatically without the help of shift engineer.

After synchronization, let us consider the two generators are operating in parallel. Now study the effects changing mechanical driving torque and excitation.

3.6 Effect of Changing Mechanical Driving Torque

In the laboratory, the mechanical driving torque can be controlled by changing the speed of the prime mover, i.e., DC motor or AC motor. In the generating stations, the mechanical driving torque of a generator can be controlled by the gate opening in case of hydro-generators and by throttle opening in case of turbo generator. For simplicity consider, cylindrical rotor synchronous generator. The results obtained are applicable to salient pole synchronous generator also. First consider generator on no load.

3.6.1 No Load Operation

Consider two generators with identical speed load characteristics running in parallel with a common terminal voltage, V_t volts.

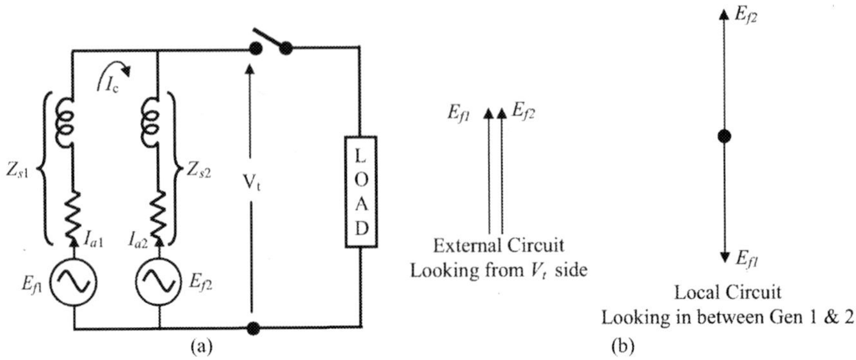

Fig. 12.10. (a) Parallel operation of generators, (b) Phasor diagrams.

Let the generated voltages of two generators be E_{f1} and E_{F2} respectively with per phase synchronous impedances be Z_{s1} and Z_{s2} respectively.

Suppose the mechanical driving torque of generator 1 is increased. Under no load condition, local circuit phasor diagram is considered. The speed of generator 1 increases and the phasor E_{f1} shifted ahead as shown in Fig. 12.11 (a). The voltage E_c appears between these two generators and current I_c circulates between two generators. The power of generator 1, $E_{f1}I_c$ cos (angle E_{f1} & I_c) is positive and power of generator 2 $E_{f2}I_c$ Cos (angle E_{f2} & I_c) is negative. Generator 1 operates as generator which resulted in decrease in speed and generator 2 operates as a motor which resulted in increase in speed. The speed of faster machine decreases and that of slower machine increases. Therefore there is auto equalization of speeds or auto synchronization action. Thus with increase in mechanical driving

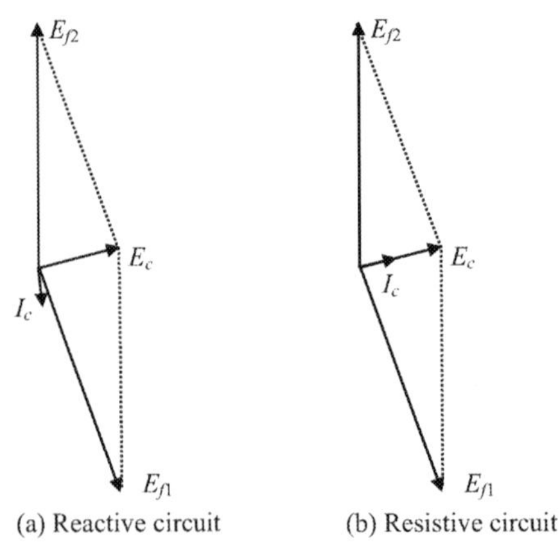

Fig. 12.11. Phasor diagram due to increased mechanical driving torque.

torque the set gets new higher speed value and with decrease in mechanical driving torque the set gets new lower speed value.

If resistive circuit is considered, the circulating current, I_c will be in phase with E_c and the power of both the generators is positive. There would not be auto synchronization action. Therefore, reactance is useful for auto synchronization action.

3.6.2 On Load Operation

When two generators are running in parallel with load impedance Z_L, the power sharing is dependent upon the speed power characteristics of prime movers and their governors. These characteristics are nearly linear and drooping as shown in Fig. 12.12.

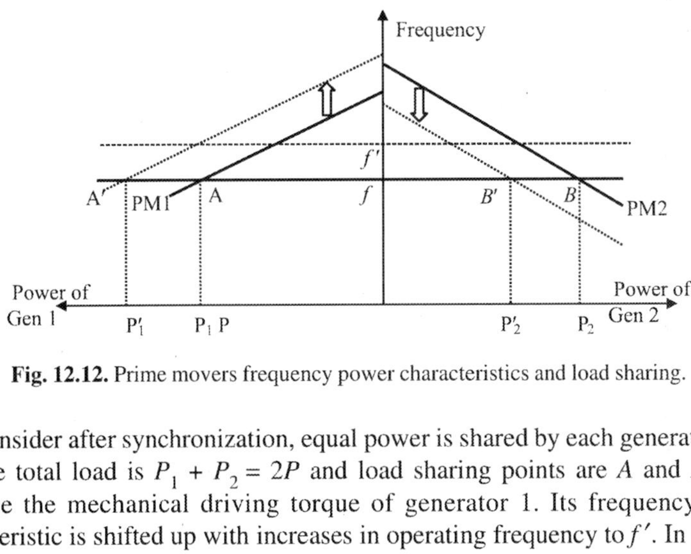

Fig. 12.12. Prime movers frequency power characteristics and load sharing.

Consider after synchronization, equal power is shared by each generator, $P_1 = P_2$. The total load is $P_1 + P_2 = 2P$ and load sharing points are A and B. Now increase the mechanical driving torque of generator 1. Its frequency power characteristic is shifted up with increases in operating frequency to f'. In order to maintain the frequency constant, the mechanical driving torque of generator 2 is reduced so that its frequency power characteristic is shifted down. Now the operating frequency is f and the load sharing points are A' and B'. With this the power shared by generator 1 is increased to $P1'$ and power shared by generator 2 is decreased to $P2'$. Again total load is $P_1' + P_2' = 2P$. Thus, the active power shared by the generator can be increased or decreased by increasing or decreasing the mechanical driving torque of that generator.

In order to study the internal behaviour of the generator, the phasor diagram is drawn with equal voltages ($E_{f1} = E_{f2}$), currents ($I_{a1} = I_{a2}$), and powers ($P_1 = P_2$).

Now increase the mechanical driving torque of generator 1. The output power increases with increase in δ to δ_1. The resultant frequency increases. In order to maintain frequency constant decrease the mechanical driving torque of generator 2. The power angle δ decreases to δ_2 with decrease in load sharing. Due to the phase difference between E_{f1} and E_{f2}, a voltage E_c appears between these two

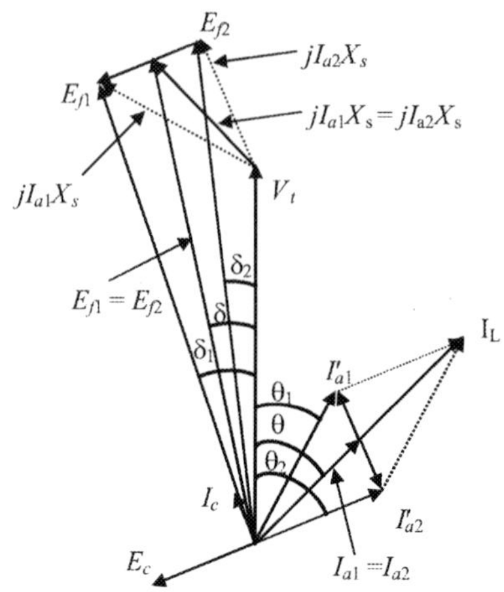

Fig. 12.13. Phasor diagram illustrating the effect of mechanical driving torque.

generators. The circulating current I_c flows. This current is added to I_{a1} and is subtracted from I_{a2}.

It is seen that the operating power factor of generator is improved if the mechanical driving torque is increased and the operating power factor of generator is worsened if the mechanical driving torque is decreased. Thus, the effect of increase in mechanical driving torque is to:

1. increase in speed and frequency
2. increase in power sharing and power angle δ
3. increase in armature current
4. improvement in power factor

3.7 Effect of Changing Excitation

In the laboratory, the field current or excitation can be controlled with the use of field rheostat. In the generating stations, the excitation of a generator can be controlled by DC exciter, static exciter and brushless excitation system. Here again the effect of changing the excitation is considered first on no load and then on load.

3.7.1 No Load Operation

For the system of two generators, again assume that two generators are identical. Their excitation voltages are equal so that that there is no current in the local circuit on no load. If the excitation of generator 1 is increased, E_{f1} increases and becomes more than E_{f2} and the voltage E_c appears between them. The circulating current *Ic* flows which lags the voltage E_c by 90° as shown in Fig. 12.14(a). The current *Ic* lags E_{f1} and leads E_{f2}. For generator 1, I_c produces demagnetizing action which resulted in lowering E_{f1}. For generator 2, I_c produces magnetizing

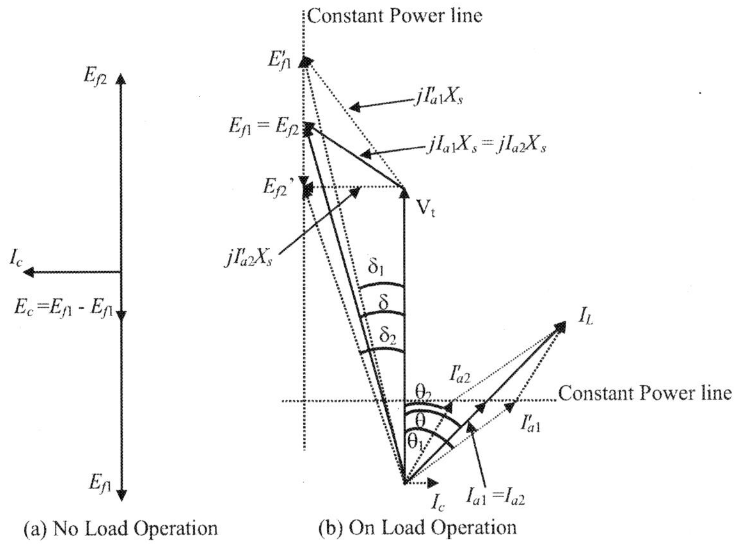

Fig. 12.14. Phasor diagrams illustrating the effect of field excitation.

action which resulted in increase in E_{f1}. The current I_c tends to lower the increased voltage E_{f1} and raise the another voltage E_{f2}. Thus, there is auto equalization of voltages due to reactance of the circuit.

Therefore with increase in excitation the set gets new higher voltage value and with decrease in excitation the set gets new lower voltage value. On no load the power angle δ remains zero.

3.7.2 On Load Operation

For identical generators the prime mover governors are adjusted for equal load sharing and their excitations are equal so that they operate with identical currents, voltages, power factors, load angles, active and reactive powers, as shown by thick line in Fig. 12.14(b). Since the function of field current is to establish flux, this will not change the active power but will change power factor and reactive power. The active power equation is

$$P = V_t I_a \cos\theta = \frac{E_f V_t}{X_s} \sin\delta \tag{2}$$

V_t and X_s are constant, therefore for constant power P,

$$I_a \cos\theta = E_f \sin\delta = \text{Constant} \tag{3}$$

The path for variation of armature current I_a gives one constant power line and path for variation of E_f gives another constant power line.

Let the excitation of generator 1 be increased. As a result E_{f1} increases to E_{f1}' and is drawn on constant power line. The terminal voltage V_t increases. In order to maintain constant voltage V_t, the excitation of generator 2 is reduced from E_{f2} increases to E_{f2}', which is again drawn on constant power line. Now due to the difference between E_{f1}' and E_{f1}', the circulating current I_c flows, which is added to I_{a1} and subtracted from I_{a2}. The I_{a1} is increased to I_{a1}' and and I_{a2} is decreased to I_{a2}'. The current I_{a1}' and I_{a2}' are drawn on constant power line. It is seen from the phasor diagram that the increase in excitation resulted to:

1. increase in voltage and decrease in power angle δ
2. increase in reactive power sharing
3. increase in armature current from unity power factor point
4. worsening the power factor

3.8 Operation of Generator when Connected to Infinite Bus

The operation of synchronous generator connected to an infinite bus is different from that when it is connected to another generator in parallel. In earlier section, it is seen that when the two generators are in parallel, a change in mechanical driving torque resulted in the change in its speed and frequency and a change in excitation resulted in the change in its terminal voltage.

But when the generator is connected to infinite bus, a change in mechanical driving torque resulted in no change in its speed and frequency and a change in excitation resulted in no change in its terminal voltage because the frequency and terminal voltage are maintained constant by infinite bus.

In this case the effect of increase in mechanical driving torque is to:

1. increase in power sharing and power angle δ
2. increase in armature current
3. improvement in power factor

And the increase in excitation resulted to:

1. increase in reactive power sharing
2. increase in armature current from unity power factor point
3. worsening the power factor

3.9 Load Sharing between Two Generators

Refer to Fig. 12.10 (a).

The no load voltage of generator 1, $\overline{E}_{f1} = \overline{V}_t + \overline{I}_{a1} Z_{s1}$ (4)

$$\therefore \overline{I}_{a1} = \frac{\overline{E}_{f1} - \overline{V}_t}{Z_{s1}}$$

The no load voltage of generator 2, $\overline{E}_{f2} = \overline{V}_t + \overline{I}_{a2} Z_{s2}$ (5)

$$\therefore \bar{I}_{a2} = \frac{\bar{E}_{f2} - \bar{V}_t}{Z_{s2}}$$

$$\bar{E}_{f1} - \bar{E}_{f2} = \bar{I}_{a1} Z_{s1} - \bar{I}_{a2} Z_{s2} \tag{6}$$

On no load, the circulating current is given by $\bar{I}_{a1} = -\bar{I}_{a2} = \bar{I}_c = \dfrac{\bar{E}_{f1} - \bar{E}_{f2}}{Z_{s1} + Z_{s2}}$ (7)

On short circuit, the current is given by $\bar{I}_{a1} = \dfrac{\bar{E}_{f1}}{Z_{s1}}$, and $\bar{I}_{a2} = \dfrac{\bar{E}_{f2}}{Z_{s2}}$ (8)

On load, the current is given by

$$\bar{I}_L = \bar{I}_{a1} + \bar{I}_{a2} = \frac{\bar{E}_{f1} - \bar{V}_t}{Z_{s1}} + \frac{\bar{E}_{f2} - \bar{V}_t}{Z_{s2}} \tag{9}$$

The terminal voltage is given by $\bar{V}_t = \bar{I}_L Z_L = \left[\dfrac{\bar{E}_{f1} - \bar{V}_t}{Z_{s1}} + \dfrac{\bar{E}_{f2} - \bar{V}_t}{Z_{s2}}\right] Z_L$

From equation 6, the current I_{a2} is given by

$$\bar{I}_{a2} = \frac{\bar{E}_{f1} - \bar{E}_{f2} + \bar{I}_{a1} Z_{s1}}{Z_{s2}} \tag{10}$$

$$\bar{E}_{f2} = \bar{I}_{a2} Z_{s2} + (\bar{I}_{a1} + \bar{I}_{a2}) Z_L \tag{11}$$

Putting equation 10 in 11, and solving for I_{al}, we get

$$\bar{I}_{a1} = \frac{\bar{E}_{f1}}{Z_{s1} + Z_L + \dfrac{Z_{s1} Z_L}{Z_{s2}}} + \frac{\bar{E}_{f1} - \bar{E}_{f2}}{Z_{s1} + Z_{s2} + \dfrac{Z_{s1} Z_{s2}}{Z_L}} \tag{12}$$

$$\bar{I}_{a2} = \frac{\bar{E}_{f2}}{Z_{s2} + Z_L + \dfrac{Z_{s2} Z_L}{Z_{s1}}} + \frac{\bar{E}_{f1} - \bar{E}_{f2}}{Z_{s1} + Z_{s2} + \dfrac{Z_{s1} Z_{s2}}{Z_L}} \tag{13}$$

The currents I_{a1} & I_{a2} in terms of load current I_L and impedances Z_{s1} & Z_{s2} are given by

$$\bar{I}_{a1} = \frac{Z_{s2}}{Z_{s2} + Z_{s2}} I_L \text{ and } \bar{I}_{a2} = \frac{Z_{s1}}{Z_{s2} + Z_{s2}} I_L \tag{14}$$

The equations of power shared by generators 1 and 2 are given by

$$P_1 = \frac{Z_{s2}}{Z_{s2} + Z_{s2}} P_L \text{ and } P_2 = \frac{Z_{s1}}{Z_{s2} + Z_{s2}} P_L \tag{15}$$

The importance and need of electric power in everyday life has reached to such a stage that everybody needs continuity of power supply. The electric power is an essential prerequisite for the progress of any country. The modern power system has features unique to itself with the present rate of growth that the power demand is almost doubling every decade. The electric power system is the tool for converting and transporting energy which is further utilized for getting useful work done from it.

The problem of stability and reliability of an interconnected power system is very complex and needs careful investigation while planning, design, layout operation and maintenance stages of power system. Stability is an important factor for satisfactory operation of a system consisting of number of generators, induction motors, synchronous motors, loads, number of transmission lines, condensers, reactors and the distribution system to connect generating stations to loads. As a result of sudden overdrawal of power and their following switching incidents, the system may go to total blackout condition. The economic scheduling, planning, operation and control of interconnected power system require a proper load flow studies of the system. In everyday operation of power system, there is possibility of disturbances in normal operation and occurrence of the faults in the different parts of the power system equipment. The most common form of disturbance in normal operation of a power system is overloading of transmission line due to which the flow of current exceeds the rated value. The most severe and common form of fault is short circuit.

In a simple power system, the equations 12, 13, 14, 15 can be used for load sharing, but in complex power system, the load sharing equation can be obtained by considering some constants K. Since the power shared by the generator is directly proportional to the bus voltage, bus voltage angle, generator power factor, and inversely proportional to the distance of the load bus from the generator bus, the load sharing equation in the general form can be written as:

$$P = K_1 + (K_2 . \text{bus voltage}) + (K_3 . \text{bus voltage angle}) + (K_4 . \textit{power factor}) + (K_5 . \textit{current}) + (K_6 . \textit{distance}) \tag{16}$$

The direct proportionality constant and inverse proportionality constant can be obtained by doing statistical regression calculations with the use of some mathematical software packages.

4. APPARATUS REQUIRED

1. Synchronous generator: Rating: (For example, 7.5 kVA, 440 V, 1500 rpm)
2. DC shunt motor (Prime mover): Rating: 5 kW, 230 V, 1600 rpm.
3. Lamp bank and synchronizing switch
4. AC ammeter: One, 0-10 amps.
5. AC voltmeter: Two, 0-300 volts.
6. Rheostat: Two, 400 Ω, 2 amps & 35 Ω, 8 amps
7. Tachometer or speedometer: One

5. CIRCUIT DIAGRAM

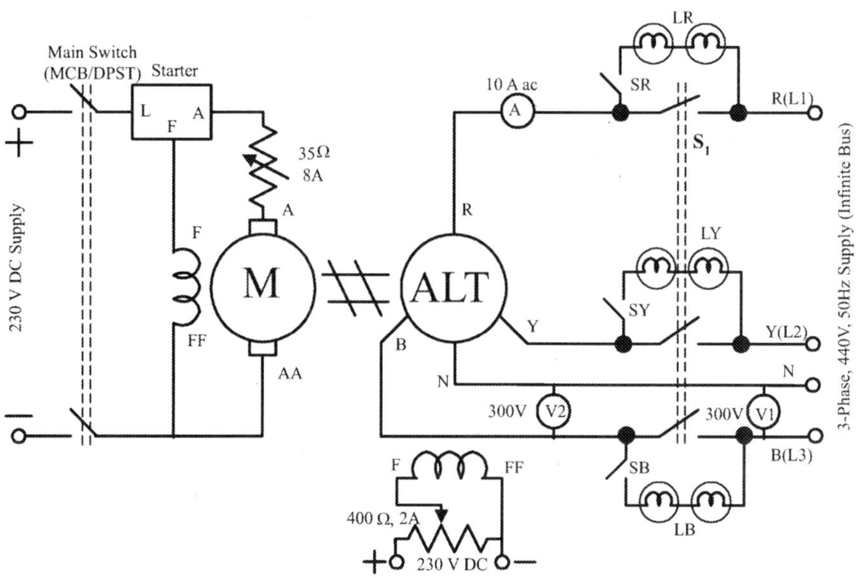

Fig. 12.15. Experimental set-up for synchronizing of an alternator with an infinite bus by the dark lamp method.

6. PROCEDURE

1. Make the connections as shown in Fig. 12.15.
2. Set the field rheostat of alternator to zero or minimum output voltage point and armature rheostat of motor to maximum resistance point.
3. Keep all the switches S_1, SR, SY and SB in open position and switch on DC supply.
4. Start DC motor and bring the speed very near to synchronous speed of the alternator.
5. Switch on AC supply from infinite bus side and measure its voltage *V*1.

6. Adjust the field rheostat of alternator such that it gives voltage *V*2 equal to V1. The error in measuring the voltage can be removed if one voltmeter is used.
7. Switch on SR, SY and SB, and watch the changes in the GLOW pattern of the three lamp banks. All the three lamps come to the fully bright position at one instant and then slowly go to dark side. The cycle repeats. If required, make small adjustment in excitation of the alternator and speed of DC motor to get long dark periods.
8. In the middle of one such dark period, CLOSE switch S_1, thus synchronizing the alternator with the infinite bus. At the time of closing the switch S_1, note the ammeter reading. It should be as small as possible.

7. OBSERVATION

Note down the reading of ammeter.

8. RESULT

The load shared by incoming generator is of .... amps.

9. CONCLUSION

An alternator can be synchronized with the infinite bus. At the time of synchronization voltage and frequency of incoming alternator should be equal to that of infinite bus. The voltage of the incoming alternator should be in phase with the infinite bus voltage with respect to the external load.

10. POST-EXPERIMENTAL QUIZ

1. Explain the reasons for doing synchronization.
2. Practically, why the incoming machine should have speed slightly more than infinite bus?
3. Explain the advantages of the bright lamp method.

CHAPTER 13

To Determine Potier Reactance of Three-Phase Alternator by "Zero Power Factor Lagging Saturation Curve"

1. OBJECTIVE

The main objectives of this experiment are:

1. to plot the open circuit saturation curve
2. to plot zero power factor lagging saturation curve and
3. to plot Potier triangle and then to calculate Potier reactance.

2. PRE-EXPERIMENTAL QUIZ

1. What do you mean by Potier reactance?
2. What do you mean by zero power factor lagging saturation curve?
3. When the armature of an alternator is shorted, the short circuit current produces magnetizing action or demagnetizing action. Explain.
4. Under short circuit conditions, even though the output is zero, why some input field current is still required?

3. THEORY

The simple model of the synchronous machine is obtained by making the assumption that the magnetic circuit of the machine is linear. But actually in practice the machine operates in the saturated region. Therefore the parameters measured under saturated condition will give more accurate results.

The Potier reactance is the leakage reactance when generator is having rated voltage and rated armature current or armature reaction at full load, i.e., generator is under saturated condition. The leakage reactance is obtained by the use of **Potier triangle**; therefore it is called Potier reactance.

3.1 Potier Triangle Method

The Potier triangle can be drawn by knowing the open circuit characteristic, the field excitation for rated armature current when alternator is short circuited and zero power factor lagging saturation characteristic.

The open circuit characteristic is the plot between excitation (I_f) and per phase open circuit voltage (E_f). The zero power factor lagging saturation curve (ZPFLSC) is a curve between field current (I_f) and armature terminal per phase voltage (V_t) when load current, speed are at rated values and power factor is zero lagging, i.e., the load is purely inductive. For an alternator, these characteristics are obtained as follows:

1. The alternator is run at rated speed by prime mover.
2. The field winding excitation is increased from zero and open circuit voltage is noted.
3. With armature shorted, field current is increased from zero in order to flow rated armature current. The field current for rated I_a is noted (only one reading).
4. A purely inductive load is connected across the armature terminals and field current is increased till full load armature current is flowing.
5. The inductive load is varied in steps and the field current at each step is adjusted to maintain rated full load armature current (ZPFLSC).

The terminal voltage V_t and air gap voltage E_r are very nearly in phase and are related by the simple algebraic equation, $V_t = E_r - I_a r_a - I_a x_{al}$. The resultant mmf F_r and the field mmf F_f are also very nearly in phase and are related by the simple algebraic equation, $F_f = F_r + F_a$.

The OCC and (FL) ZPFLSC are shown in Fig. 13.1. For field excitation F_f (I_f), equal to OH, the open circuit voltage is *HF* (E_f). The curves are on per phase basis. OH is also the excitation required to produce rated terminal voltage at full load with zero pf lagging and *OA* is the excitation required to get rated armature current when terminals of alternator are short circuited.

Under short circuit condition, the terminal voltage V_t is zero and the phasor diagram is shown in Fig. 13.2. The air gap emf E_r, generated by the resultant mmf F_r or air gap flux Φ_r, is sufficient to overcome the **leakage impedance drop $I_a(r_a + jx_{al})$**. It can be expressed as $\bar{E}_r = \bar{I}_a r_a + j\bar{I}_a x_{al}$. Generally **$x_{al}$ is much greater than r_a** and, therefore, the armature **current I_a lags E_r by an angle of nearly 90° (inductive winding)**. Phasor diagram reveals that F_a is almost opposite to F_f, so F_a is almost entirely demagnetizing in nature. The air gap flux is reduced showing no saturation under short circuit condition.

The point *A* in Fig. 13.1 (b), on ZPFLSC corresponds to the zero terminal voltage and can, therefore, be obtained from short circuit test. The excitation OA can be divided into two parts OC and CA. The field due to armature reaction is in direct opposition to the main field because **power factor is zero lagging** as shown

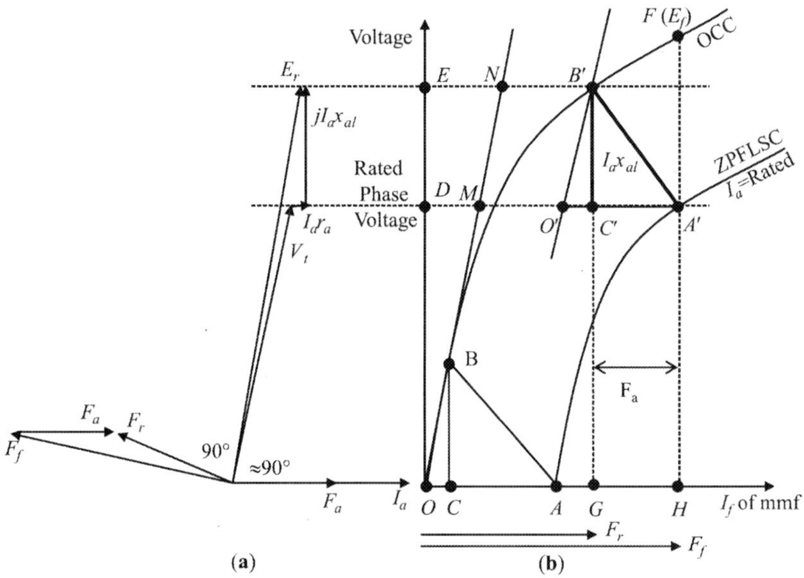

Fig. 13. 1. (a) Phasor diagram of cylindrical rotor alternator at zero power factor overexcited. (b) OCC, SCC, ZPFLSC and potier triangle.

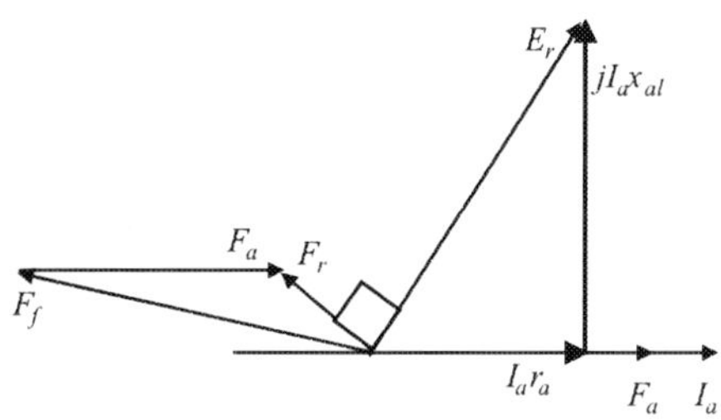

Fig. 13.2. Phasor diagram of the alternator for short circuit condition.

in Fig. 13.1 (a). Due to this phase opposition, more field current CA is required to overcome $I_a r_a$ or to counter balance the armature reaction (i.e., armature mmf).

The remaining excitation OC generates the voltage which is equal to the leakage reactance drop $I_a x_{al}$, which is shown by vertical line BC in Fig. 1 (b). However, the present position of C is assumed and is not known.

The rated phase voltage line DA' is drawn. The point A' is on ZPFLSC and mmf is F_f, which is also responsible for no load voltage $E_f = HF$. With the known value of OA (from short circuit test), cut the line DA' at O' so that $OA = O'A'$. From point O', draw a line parallel to air gap line, cutting OCC at B', which gives air gap voltage E_r, due to resultant mmf F_r. The same mmf F_r is responsible for the generation of terminal voltage V_t. From B', a perpendicular on DA', gives point C' and leakage reactance voltage drop $B'C'$. The armature leakage reactance $x_{al} =$

(B'C'/I_a rated) = x_p. The difference between F_f and F_r is F_a, which is equal to C'A'.

The triangle obtained after joining the points O', A' and B', is called the Potier triangle. Thus from Potier triangle, the armature leakage reactance x_{al} and armature reaction mmf F_a can be determined.

If the armature resistance is assumed zero and the armature current is kept constant, then the size of Potier triangle O'A'B' remains constant and can be shifted parallel to itself with its corner B' remaining on OCC and corner A' tracing the ZPFLSC. Under ZPFL condition, the armature current is having rated value and the machine is under saturated condition. Due to more leakage flux under saturated condition, more field current is required to produce same air gap flux. Therefore, the ZPFLSC has the same shape as the OCC but at lower level. With respect to OCC the ZPFLSC is shifted downward by an amout equal to leakage reactance voltage drop and horizontally to the right by an amount equal to armature reaction mmf. The leakage reactance obtained by Potier triangle is slightly more than the actual armature leakage reactance but for all practical purposes, the Potier reactance is taken as armature leakage reactance.

The ZPFLSC can also be obtained by loading over-excited generator (lagging) by under excited synchronous motor (lagging). The field current of these two machines are varied in such a way that the generator armature current is kept constant to its rated value.

The ZPFLSC is a load characteristic. If the power factor in increased from zero to one, the opposition to the field mmf due to armature current decreases and the curve shifts towards OCC as shown in Fig. 13.3.

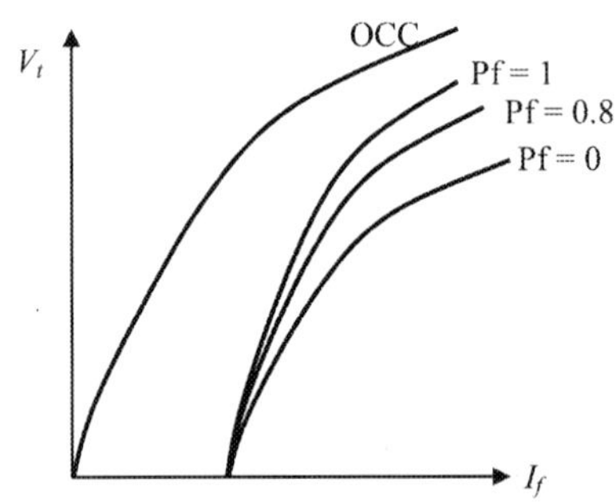

Fig. 13.3. Load characteristic of a generator with variable power factor (lagging) and constant armature current.

3.2 Compounding Curves for a Synchronous Generator

Depending on the opposition to the field mmf due to armature current, the compounding curves of a synchronous generator can be drawn, as shown in Fig. 13.4. For a particular constant voltage, the requirement of field current decreases as the power factor changes from ZPF lagging to ZPF leading. This is a useful

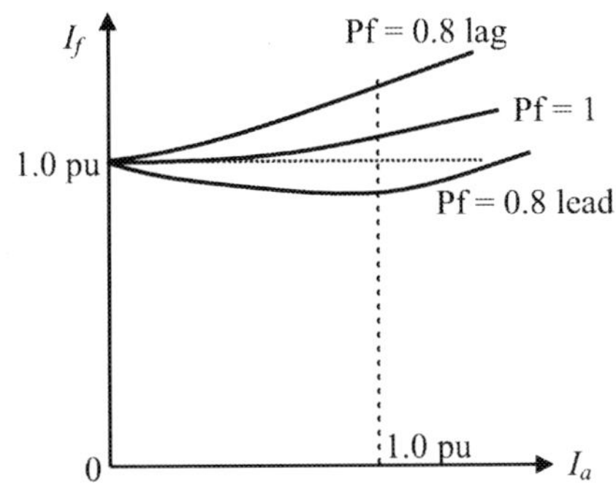

Fig. 13.4. Compounding curves for a synchronous generator.

characteristic as a guide to the engineer working at generating station as far as generator operation is considered. Thus, it can be concluded that for lagging power factor, the terminal voltage decreases and for leading power factor, the voltage increases.

3.3 Voltage Regulation

3.3.1 ZPFLSC Method

This method is also useful for calculating voltage regulation. For any load, first E_r is calculated by using the relation $\bar{E}_r = \bar{V}_t + \bar{I}_a(r_a + jx_p)$. According to the magnitude of E_r, F_r is obtained from OCC and is drawn leading by 90° to E_r. By knowing F_a from Potier triangle, F_f is calculated by using the relation $\bar{F}_f = \bar{F}_r + \bar{F}_a$.

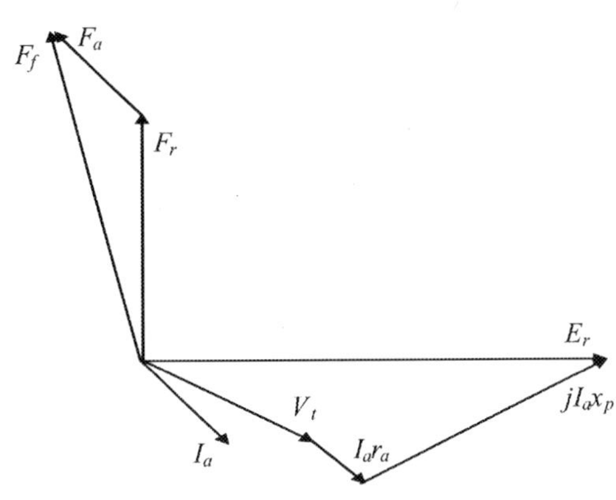

Fig. 13.5. Phasor diagram for determination of voltage regulation by ZPFLSC method.

Corresponding to F_f, no load voltage E_f is obtained from OCC and then voltage regulation is calculated by using the relation % regulation $=\{(E_f - V_t)/V_t\}\times 100$.

3.3.2 EMF Method or Synchronous Impedance Method

By knowing the values of resistance r_a and synchronous reactance X_s from OCC and SCC, the phasor diagram for any load and any power factor can be easily drawn, as shown in Fig. 13.6. Since the phasor diagram is drawn with the use of emf and is used for the calculation of voltage regulation; this method is called the emf method. The reactance is constant with respect to the air gap position in cylindrical rotor machine; therefore this method is suitable for cylindrical rotor generator.

From the phasor diagram, we can write

$$\bar{E}_f = \bar{V}_t + \bar{I}_a Z_s$$

$$= V_t + j0 + (I_a\cos\theta - jI_a\sin\theta)(r_a + jX_s)$$

$$= (V_t + I_a r_a\cos\theta + I_a X_s\sin\theta) + j(I_a X_s\cos\theta - I_a r_a\sin\theta) \quad (1)$$

$$E_f = \sqrt{(V_t + I_a r_a\cos\theta + I_a X_s\sin\theta)^2 + j(I_a X_s\cos\theta - I_a r_a\sin\theta)^2} \quad (2)$$

The percentage voltage regulation is given by

$$\%R = \frac{E_f - V_t}{V_t}\times 100\% \quad (3)$$

The regulation can also be calculated for leading power factor by putting θ as negative and for unity power factor by putting θ as zero.

The impedance under unsaturated region is more than that under saturated region. The impedance obtained from OCC and SCC is having higher value since the machine is under short circuit test and is operating in unsaturated region. The voltage regulation calculated by the emf method is having higher value than the actual value. Therefore this method is called ***pessimistic*** method.

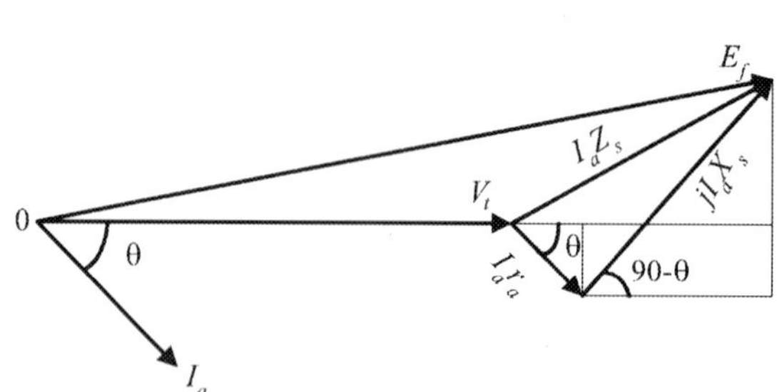

Fig. 13.6. Phasor diagram: EMF method.

3.3.3 Ampere Turn or MMF Method

In this method the data of OCC and SCC is used. The mmf phasor diagram is drawn as shown in Fig. 13.7. From $\overline{V}_t + \overline{I}_a r_a$, F_{r1} is obtained with the help of OCC. From SCC, $F_a + F_{al}$ is obtained. Depending on the power factor, $\overline{F}_f = \overline{F}_{r1} + (\overline{F}_a + \overline{F}_{al})$ is obtained. Now corresponding to field mmf F_f, E_f is obtained from OCC and thus the percentage regulation can be calculated from the relation

$$\% R = \frac{E_f - V_t}{V_t} \times 100$$

The emf and mmf methods are based on the assumption of an unsaturated magnetic circuit and since all the machines are operated under saturation condition, these methods cannot give reliable results. If OCC is considered a straight line characteristic, similar to air gap line, then all the three methods will give the same results.

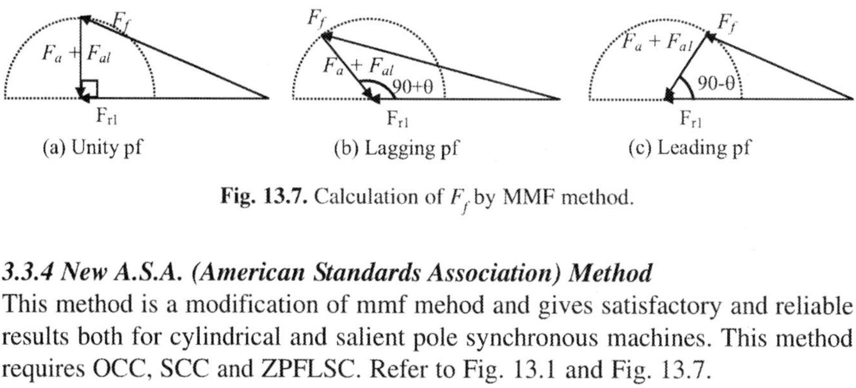

Fig. 13.7. Calculation of F_f by MMF method.

3.3.4 New A.S.A. (American Standards Association) Method

This method is a modification of mmf mehod and gives satisfactory and reliable results both for cylindrical and salient pole synchronous machines. This method requires OCC, SCC and ZPFLSC. Refer to Fig. 13.1 and Fig. 13.7.

For rated voltage V_t, F_{r1} is taken equal to DM depending on air gap line. The line F_a + F_{al} is drawn from the SC test. The resultant F_f is calculated.

By knowing the resistance and reactance, air gap voltage E_r is calculated as per equation $\overline{E}_r = \overline{V}_t + \overline{I}_a(r_a + x_{al})$. The horizontal line for E_r is drawn crossing at *N* with air gap line and at *B*' with OCC. The length *NB*' is added to F_f in order to get the total excitation. With respect to this total excitation, the excitation voltage E_f is obtained from OCC and the regulation is calculated.

This method is used in the USA.

4. APPARATUS REQUIRED

1. Synchronous machine: Rating: (For example, 7.5 kVA, 440 V, 1500 rpm)
2. DC shunt motor (prime mover): Rating: 5 kW, 230 V, 1600 rpm
3. AC Ammeter: Three, 0-10 Amps.
4. AC Voltmeter: One, 0-300 Volts.
5. DC Ammeter: One 0-1 Amp.
6. Rheostat: Two, 400 Ω, 2 Amps and 300 Ω, 2 Amps.
7. Variable inductor: Three, 230 V, 24 A,
8. Triple pole double throw switch: One.
9. Tachometer or speedometer: One

5. CIRCUIT DIAGRAM

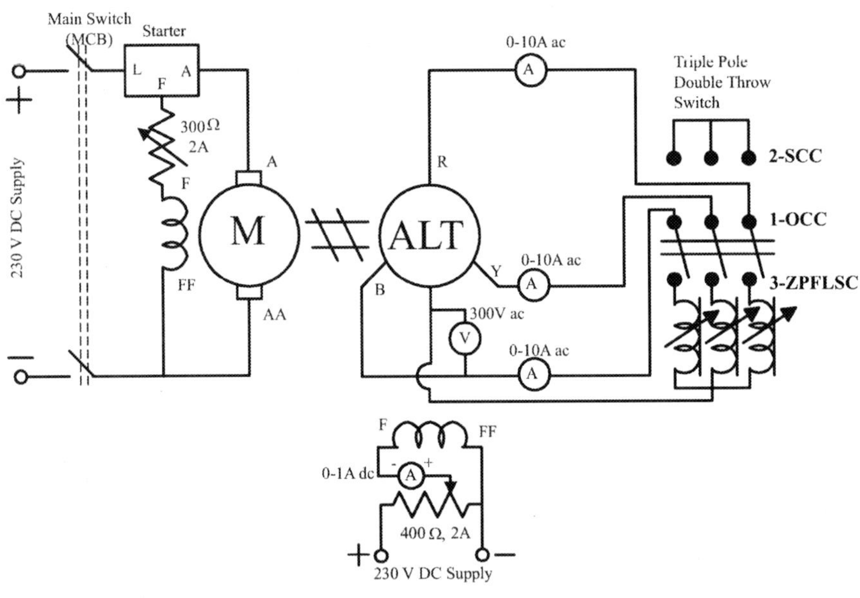

Fig. 13.8. Experimental set-up for determination of Potier reactance.

6. PROCEDURE

1. Make the connections as shown in Fig. 13.8.
2. Set potential divider (connected to field winding of alternator) to zero output position and triple pole double throw switch to open position, i.e., position no. 1 in which alternator terminals are all open.
3. Set field rheostat of motor to minimum resistance point or set the speed control handle to slow speed point (if available in starter of DC motor).
4. Switch on DC supply and start DC motor with the help of starter.

5. Adjust the motor speed to synchronous speed of alternator with the help of field rheostat or speed control handle.
6. Note voltmeter reading. Increase the excitation in step and note the corresponding voltage reading. Take reading up to 20% above the rated voltage.
7. Bring the potential divider to zero output position and close the **TPDT switch to position no 2,** in which alternator terminals are shorted.
8. Increase the excitation till ammeters read **rated current of alternator**. Note the value of field current. This value gives starting point of ZPFLSC i.e., the position of point A in Fig. 13. 1.
9. **Bring the potential divider to zero output position** and close the TPDT switch to position no 3, in which RYB of alternator are connected to three variable inductors.
10. Keep variable inductors to minimum reactance position (Top) and increase excitation such that all the ammeters read rated armature current of alternator. Note field current and the armature voltage.
11. Now increase the excitation in step and increase reactance of inductors (towards bottom) such that the ammeters again read rated armature current of alternator. Note field current and the armature voltage.
12. Repeat step no. 11 to get readings up to rated armature voltage.
13. Reduce the excitation to zero and switch off the DC supply.

7. OBSERVATIONS

Speed of the set = Synchronous speed = ... rpm.

1. Open Circuit Test

Fielf Current I_f A	Arm Terminal Voltage/phase, V_1
0.1 A	
0.2 A	
..	
...	
...	20% above related value

2. Short Circuit Test

Field current required for getting rated armature current of alternator, I_f = ...Amp

3. Zero Power Factor Lagging Saturation Test

Rated armature current of alternator, I_a = ...Amp

Fielf Current $I_f A$	Arm Terminal Voltage/phase, V_1	Arm current $I_a A$
		Rated
	up to rated value	

8. RESULT

Plot OCC, ZPFLSC (Fig. 13.1(b)) and from this draw Potier triangle. Measure *B'C'*.

$$x_{al} = x_p = \frac{B'C'}{\text{Rated armature current}}$$

9. CONCLUSION

The Potier reactance x_p of the alternator is found to be ......Ω i.e., ...per unit

10. POST-EXPERIMENTAL QUIZ

1. Under short circuit condition, why field current required to overcome resistive drop is less than that of inductive drop?
2. How Potier reactance is useful for determining voltage regulation?
3. Give the list of other methods used for determining voltage regulation and compare them.

CHAPTER 14

To Determine Negative Sequence and Zero Sequence Reactances of Synchronous Generator

1. OBJECTIVE

The main objectives of this experiment are:

1. to determine negative sequence reactance and
2. to determine negative sequence reactance of a synchronous machine.

2. PRE-EXPERIMENTAL QUIZ

1. What do you mean by negative sequence reactance?
2. Which harmonic is responsible for negative sequence component?
3. What do you mean by zero sequence reactance?
4. Which harmonic is responsible for zero sequence component?

A. NEGATIVE SEQUENCE REACTANCE

3. THEORY

3.1 Symmetrical Components

A set of three phasors forming a three-phase unbalanced system (i.e., unbalanced load on three-phase synchronous generator) can be resolved into their symmetrical components by considering sequence networks of the generator.

1. A set which consists of three-phasors which are equal in magnitude and are displaced by 120° in phase from each other and have the ***same*** phase sequence as the original phasors. These are known as positive sequence components. These components are same as that of balanced system.
2. A set consisting of three phasors which are equal in magnitude and are displaced by 120° in phase from each other and have the phase sequence ***opposite*** to that of the original phasors. These are known as **negative sequence components.**

3. A set consisting of three phasors which are equal in magnitude and have ***no phase*** displacement, i.e., all of three are in phase with each other. These are known as **zero sequence components.**

The presence of positive sequence currents in the armature of a synchronous machine produce armature mmf which rotates at synchronous speed in the same direction as that of the field. As such the armature mmf and field mmf are stationary with respect to each other and have no induction effect under steady state.

The armature mmf wave actually consists of a space fundamental component and many other space harmonic components. For a three-phase system, the voltage equations with harmonic can be written as

$$\left.\begin{aligned} V_R &= V_{m1}\sin\omega t + V_{m3}\sin 3\omega t + V_{m5}\sin 5\omega t + \dots\dots\dots \\ V_Y &= V_{m1}\sin(\omega t - 120) + V_{m3}\sin 3(\omega t - 120) + V_{m5}\sin 5(\omega t - 120) + \\ V_B &= V_{m1}\sin(\omega t - 240) + V_{m3}\sin 3(\omega t - 240) + V_{m5}\sin 5(\omega t - 240) + \end{aligned}\right\} \quad (1)$$

Here V_{m1}, V_{m3}, V_{m5} are the maximum values of the fundamental, third and fifth harmonic components of the applied voltage, respectively.

The fundamental component voltages are:

$$\left.\begin{aligned} V_{R1} &= V_{m1}\sin\omega t && \text{for phase R} \\ V_{Y1} &= V_{m1}\sin(\omega t - 120) && \text{for phase Y} \\ V_{B1} &= V_{m1}\sin(\omega t - 240) && \text{for phase B} \end{aligned}\right\} \quad (2)$$

These voltages form a three-phase balanced system with phase sequence RYB as shown in Fig. 14.1.

The fifth harmonic voltages are:

$$\left.\begin{aligned} V_{R5} &= V_{m5}\sin 5\omega t && \text{for phase R} \\ V_{Y5} &= V_{m5}\sin(5\omega t - 600) = V_{m5}\sin(5\omega t - 240) && \text{for phase Y} \\ V_{B5} &= V_{m5}\sin(5\omega t - 1200) = V_{m5}\sin(5\omega t - 120) && \text{for phase B} \end{aligned}\right\} \quad (3)$$

It is seen from the above equations that the fifth harmonic voltages are having time phase displacement of 120° with phase sequence RBY opposite to fundamental

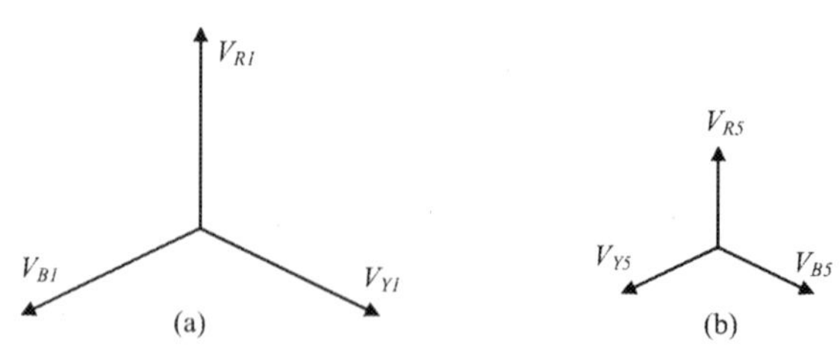

Fig. 14.1. Phasor diagrams for fundamental and fifth harmonic voltages.

components, as shown in Fig. 14.1(b). Therefore the fifth harmonic components help in producing the negative sequence reactance at its synchronous speed.

3.2 Definition

For defining, negative sequence reactance, X_2, the unexcited field winding is closed and is assumed on the rotor. The presence of negative sequence currents in the armature produce armature mmf rotating in a direction opposite to that of the field or rotor. With respect to field, it rotates at double the synchronous speed in the direction opposite to normal direction of rotation. It will induce a voltage in the damper bars as well as in field at double the line frequency. The rotating armature mmf comes in line with d-axis and soon after **{=*(15/PN)sec*}** it gets aligned with q-axis and so on. Since the flux linkages with rotor circuit are zero, these must remain zero under the conditions. Consequently, the flux paths are mainly through the air gap, similar to sub-transient reactances X_d'' and X_q''.

Under these conditions, the armature flux linkages per armature ampere, is called negative sequence inductance L_2 and $X_2 = \omega L_2$. Since the armature mmf wave experience d-axis reluctance corresponding to X_d'' and q-axis reluctance corresponding to X_q'', it is usual to take X_2 as the arithmetic mean of X_d'' and X_q'', i.e., $X_2 = (X_d'' + X_q'')/2$. As X_q'' is slightly higher than X_d'', the value X_2 does not differ much from the positive sequences reactances X_d'' and X_q''. The value of X_2, however, is less than the other positive sequence reactances of a synchronous machine.

The negative sequence network model of a synchronous machine, on three-phase and single-phase basis are shown in Fig. 14.2. The reference bus is at neutral potential which is the same as ground potential. The negative sequence voltage of phase a with respect to reference bus is

$$V_{a2} = -Z_2 I_{R2}.$$

The negative sequence component appears only the system is unbalanced and is due to fifth harmonic component of armature mmf wave. The extreme case of unbalance is a fault, line to ground, line to line or line to line to ground. In case of

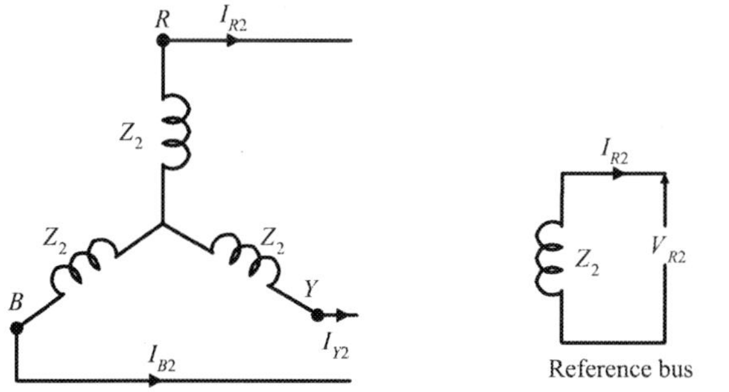

(a) Three-phase model (b) Single-phase model

Fig. 14.2. Negative sequence network of a synchronous machine.

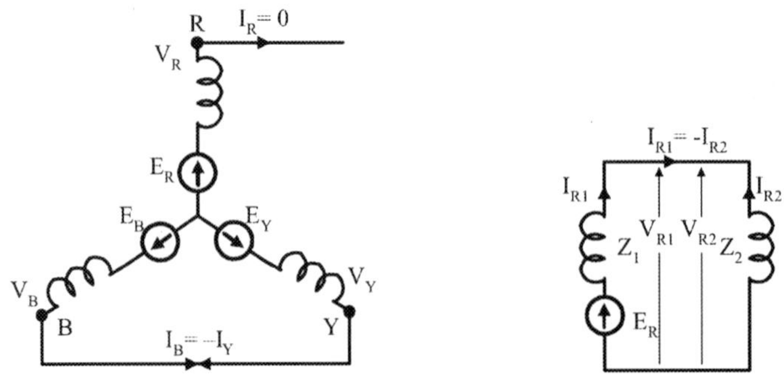

(a) Circuit diagram (b) Sequence network

Fig. 14.3. Line to line fault of unloaded generator.

line to line fault on an unloaded synchronous generator, the sequence network can be drawn in Fig. 14.3. The condition at the fault are $V_Y = V_B$, $I_R = 0$ and $I_Y = -I_B$.

Using basic equations for sequence components and substituting the above, we get

$I_{R0} = (1/3)(I_R + I_Y + I_B) = 0$, $I_{R1} = (1/3)(I_R + aI_Y + a^2 I_B)$ and

$I_{R2} = (1/3)(I_R + a^2 I_Y + aI_B)$, so $I_{R1} = -I_{R2}$

Similarly, $V_{R1} = (1/3)(V_R + aV_Y + a^2 V_B)$ and $V_{R2} = (1/3)(V_R + a^2 V_Y + aV_B)$, so $V_{R1} = V_{R2}$. Depending on this, Fig. 14.3(b) is drawn. From this Fig., $V_{R1} = E_R - I_{R1} Z_1 = I_{R1} Z_2 = V_{R2}$. Solving all the above equations for Z_2, the expression for Z_2 can be written as $Z_2 = V_{RY} / \sqrt{3}\, I_B$ (Refer Appendix).

3.3 Measurement Methods

This reactance can be measured by three different methods.

3.3.1 Measuring Sub-transient Reactances

The negative sequence reactance per phase can be calculated by

$$X_2 = (X_d'' + X_q'')/2 \tag{4}$$

3.3.2 Definition Method

The second method follows directly from the **definition of X_2**. In this method, the field winding (on rotor) is short circuited and the rotor is driven clockwise. Balanced three-phase voltages are applied to armature so that constant amplitude armature mmf rotates anticlockwise. The applied voltage is adjusted and the readings of voltmeter, ammeter and wattmeter are recorded.

Then, negative sequence impedance is $Z_2 = (V/\sqrt{3})/I$ and negative sequence reactance per phase is $X_2 = Z_2 \sin\theta$.

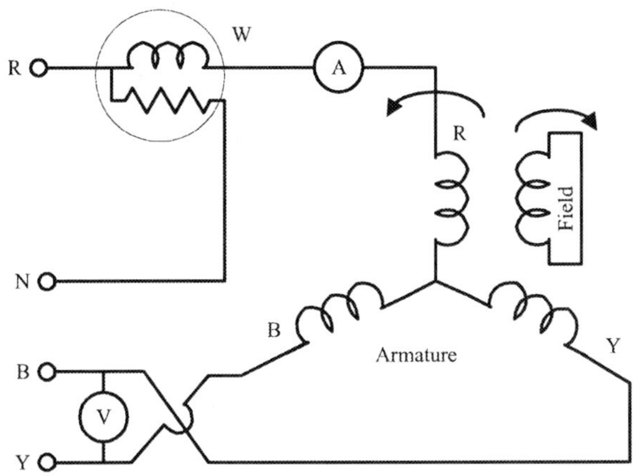

Fig. 14.4. Measurement of X_2 as per definition

$$X_2 = \frac{V}{\sqrt{3}I}\sqrt{1-\frac{P^2}{\left(\frac{V}{\sqrt{3}}I\right)^2}} = \frac{V}{\sqrt{3}I}\sqrt{1-\frac{P^2}{\left(\frac{V^2I^2}{3}\right)}} = \frac{1}{I^2}\sqrt{\frac{V^2I^2}{3}-P^2} \tag{5}$$

If negative sequence resistance is neglected, then $X_2 = (V/\sqrt{3})/I$ (6)

3.3.3 Unbalanced Fault Method

In the third method, the negative sequence impedance can be measured by referring to the diagram of Fig. 14.4, i.e., making the line to line fault. The two phases are short circuited through an ammeter and a current coil of a wattmeter as shown in Fig. 14.5. The rotor is driven at synchronous speed, with field winding unexcited and short circuited. Single phase voltage is applied and meter readings are recorded from which the negative sequence impedance is $Z_2 = V/\sqrt{3}\mathrm{I}$.

With wattmeter reading P, it can be shown that negative sequence reactance is given by

$$X_2 = \frac{V}{\sqrt{3}I}\times\frac{P}{VI} = \frac{P}{\sqrt{3}I^2} \tag{7}$$

During this test, the short-circuit current should be kept low in order to avoid undue heating of the field system.

The value of negative sequence reactance varies from 0.15 to 0.35 in per unit.

4. APPARATUS REQUIRED

1. Synchronous generator: Rating: (For example: 7.5 kVA, 440 V, 1500 rpm)

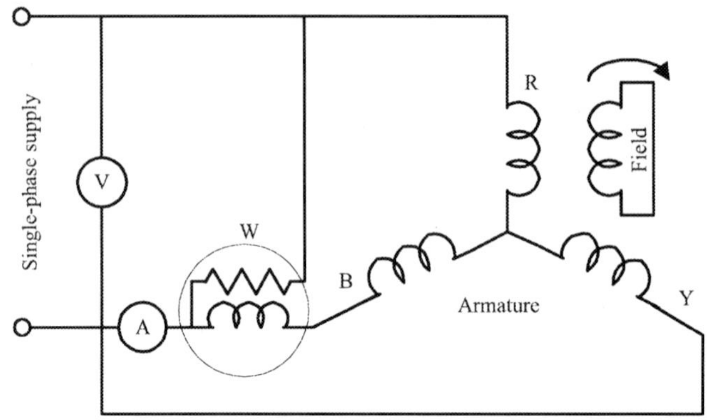

Fig. 14.5. Measurement of X_2 by making line to line fault.

2. DC shunt motor (Prime mover): Rating: 5 kW, 230 V, 1600 rpm
3. AC ammeter: One, 0-10 Amps.
4. AC voltmeter: One, 0-150-300 Volts.
5. Wattmeter: One, 300 V, 10A
6. DC ammeter: One 0-1 Amp
7. Rheostat: Two, 400 Ω, 2Amps and 35 Ω, 10 Amps.
8. Single pole switches: Two.
9. Tachometer or speedometer: One

5. CIRCUIT DIAGRAM

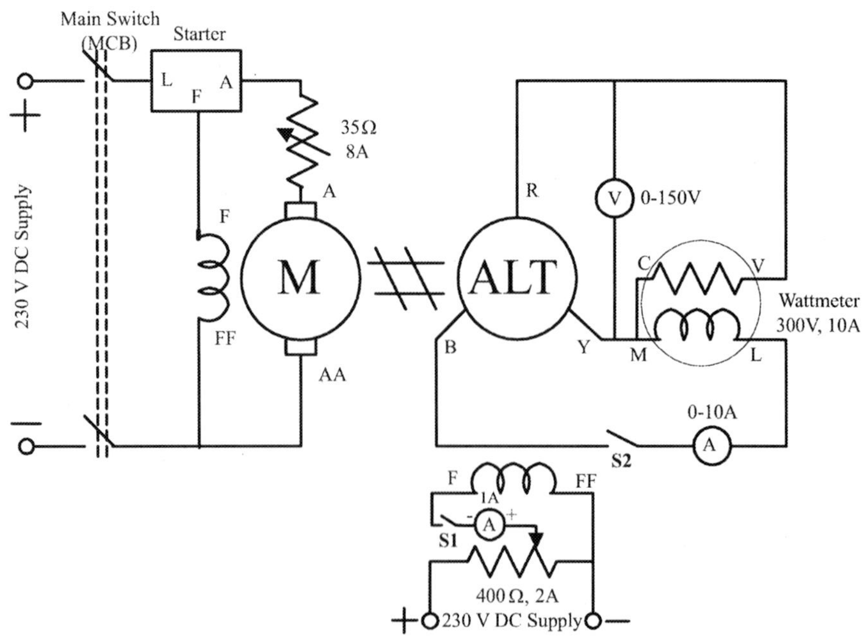

Fig. 14.6. Experimental set up for determination of negative sequence reactance.

6. PROCEDURE

1. Make the connections as shown in Fig. 14.6.
2. Set potential divider (connected to field winding of alternator) to zero output position and armature rheostat of DC shunt motor to maximum resistance point.
3. Keep single pole switches S1 and S2 open.
4. Switch on DC supply and start DC motor with the help of starter.
5. Adjust the motor speed to synchronous speed of alternator with the help of armature rheostat.
6. Close switch S1 and confirm minimum excitation current.
7. After confirming, close switch S2.
8. Increase the alternator excitation in order to flow armature current less than rated current.
9. Check the speed. It should be synchronous. If not, adjust it to synchronous.
10. Note the readings of ammeter, voltmeter and wattmeter.
11. Repeat step no. 8 to flow rated armature current.
12. Repeat step no. 9.
13. Repeat step no. 10.

7. OBSERVATION AND CALCULATIONS

Observation (for 6A/ph rting)				Calculations		
				Negative seq. impedance	Negative seq. reactance	Average X_2
Current *Isc A*	Speed	Voltage $V_{RY}\,V$	Power P	$Z_2 = \dfrac{V_{RY}}{\sqrt{3}I_{SC}}$	$X_2 = Z_2\left\{\dfrac{P}{V_{RY}I_{SC}}\right\}$	
4 A	Synchronous					
5 A						
6 A						

B. ZERO SEQUENCE REACTANCE, X_0

3. THEORY

From equation 1, the third harmonic voltages are:

$$\left.\begin{aligned} V_{R3} &= V_{m3}\sin 3\omega t, \text{ for phase R} \\ V_{Y3} &= V_{m3}\sin(3\omega t - 360) = V_{m3}\sin 3\omega t, \text{ for phase } Y \\ V_{B3} &= V_{m3}\sin(3\omega t - 720) = V_{m3}\sin 3\omega t, \text{ for phase } B \end{aligned}\right\} \quad (8)$$

It is seen from the above equations that the third harmonic voltages are in time phase with displacement of 0°, as shown in Fig. 14.7. Therefore, the third harmonic components help in producing the zero sequence reactance at its synchronous speed.

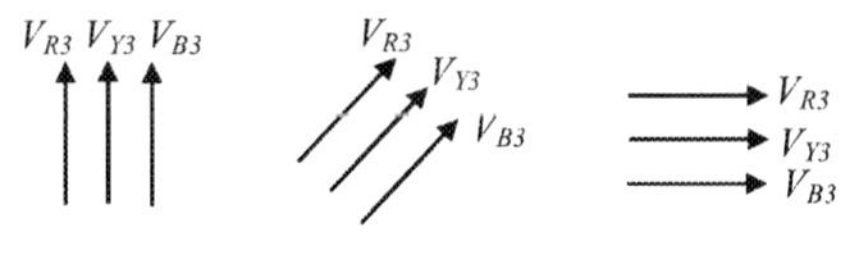

Fig. 14.7. Phasor diagrams for third harmonic voltages.

3.1 Definition

A set consisting of three phasors which are equal in magnitude and have no phase displacement, i.e., all of three are in phase with each other. These are known as zero sequence components.

The voltage and current of a three-phase unbalanced system can be obtained by phasor addition of positive sequence, negative sequence and zero sequence voltages and currents. Thus, it is clear that the zero sequence component appears only when the system is unbalanced. The extreme case of unbalance is a fault, line to ground, line to line or line to line to ground.

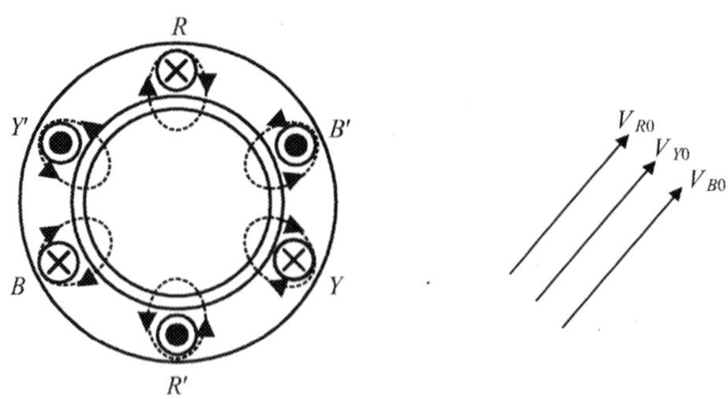

(a) Flux paths (b) Zero sequence components

Fig. 14.8. Zero sequence reactance x_0.

When armature mmf wave is not rotating but pulsating, the flux completes its path through nearby areas as shown in Fig. 14.8. Under this condition, the armature flux linkages per armature ampere is called zero sequence inductance L_0 and the reactance associated with armature is zero sequence reactance $x_0 = \omega L_0$. Zero sequence reactance is due to the third and triplen harmonic component of armature mmf wave.

3.2 Zero Sequence Network

The zero sequence network models of a synchronous machine, on three-phase and single-phase basis are shown in Fig. 14.9. The current flowing in the impedance Z_n between neutral and ground is $I_n = 3I_{R0}$. The zero sequence voltage of the terminal R with respect to ground (reference bus) is

$$V_{R0} = -3Z_n I_{R0} - Z_{g0} I_{R0} = -(3Z_n - Z_{g0}) I_{R0} = -Z_0 I_{R0} \quad (9)$$

where Z_{g0} is the zero sequence impedance per phase of the machine. Z_0 is the total zero sequence impedance of the single-phase network.

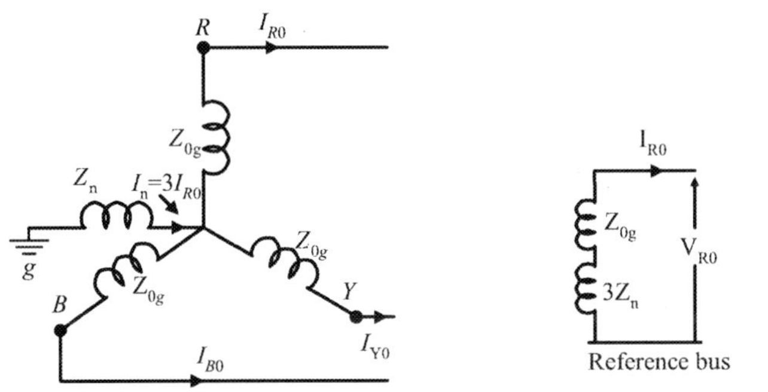

(a) Three-phase model (b) Single-phase model

Fig. 14.9. Zero sequence network of a synchronous machine.

3.3 Measurement Methods

If the three-phase armature winding is appropriately connected in series or in parallel and a single phase low voltage supply is given to these circuits as shown in Fig. 14.9, then the armature leakage flux is produced as illustrated in Fig. 14.7 (a) for a full pitch coil. The armature leakage flux is stationary. This pulsating armature mmf is opposed by currents induced on the field structure. Consequently the leakage air gap flux produced is very little.

In Fig. 14.10, field winding is short circuited and **rotor is driven at synchronous speed**. If V and I are the voltmeter and ammeter readings respectively, then for series connection, $x_0 = \dfrac{V/3}{I}\Omega$ and in for parallel connection $x_0 = \dfrac{V}{I/3}\Omega$.

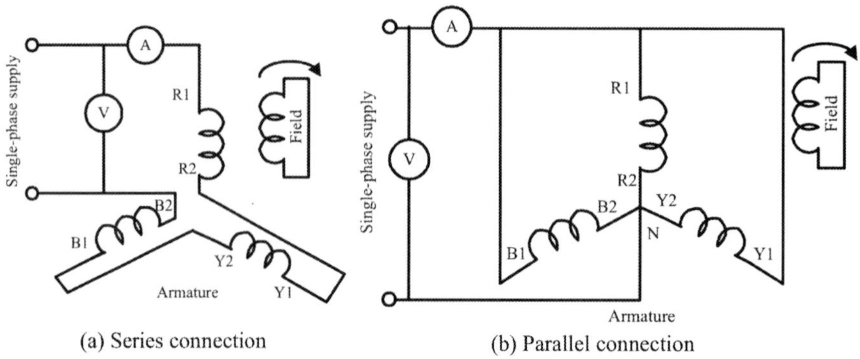

Fig. 14.10. Measurement of x_0.

The zero sequence reactance x_0 is the smallest of all the synchronous machine reactances and its value varies from 0.04 to 0.1 in per unit.

Since the flux is very less for low voltage, the same results will be obtained with rotor stationary or rotor entirely removed.

4. APPARATUS REQUIRED

1. Synchronous generator: Rating: (For example, 7.5 kVA, 440V, 1500 rpm)
2. DC shunt motor (Prime mover): Rating: 5 kW, 230 V, 1600 rpm.
3. Dimmerstat: One, 20 A, input 230 V, output 230 V
4. Ammeter: One, 0-20 Amps.
5. AC voltmeter: One, 0-30 volts.
6. Tachometer or speedometer: One

5. CIRCUIT DIAGRAM

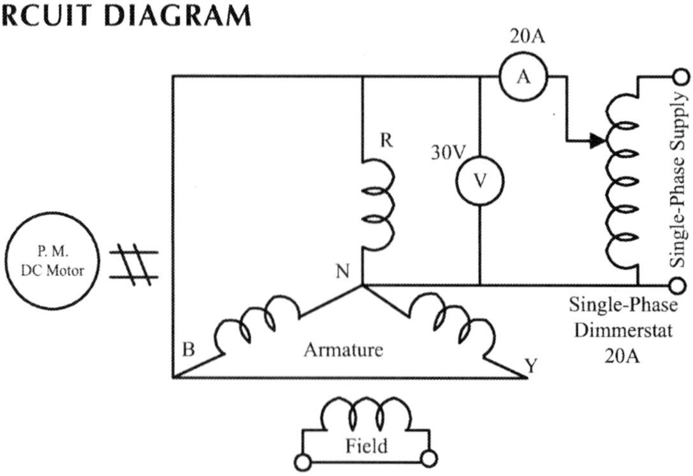

Fig. 14.11. Experimental set-up to determine zero sequence reactance.

6. PROCEDURE

1. Make the connections as shown in Fig. 14.11.
2. Run the generator at synchronous speed.
3. Set the dimmerstat output to zero volt and switch on single-phase supply.
4. Gradually increase dimmerstat output such that current through armature is less than the rated current.
5. Note down ammeter and voltmeter readings.
6. Repeat step no. 4 upto the rated armature current.
7. Repeat step no. 5.

7. OBSERVATION AND CALCULATIONS

Observation (for I_a = 6A/ph rating)			Calculations	
Speed	Current A	Voltage V	$x_0 = \frac{3V}{I}\Omega$	Average x_0
Synchronous	12 A			
	14 A			
	18 A			

8. RESULT

The negative sequence and zero sequence reactances of an alternator are determined and are found to be 1. X_2=...... Ω i.e., ...per unit. 2. x_0 =...... Ω i.e., ...per unit.

9. POST-EXPERIMENTAL QUIZ

1. Why negative sequence reactance is equal to the average of X_d'' and X_q''.
2. In the measurement of negative sequence reactance, why speed decreases when fault current is increased?
3. In the measurement of negative sequence reactance, which type of power is measured by wattmeter?
4. Compare zero sequence reactance and leakage reactance.

Appendix

Operator

$$a = 1\angle 120 = 1e^{j2\pi/3} = -0.5 + j0.866, \qquad a^2 = 1\angle 240 = 1e^{j4\pi/3} = -0.5 - j0.866$$

$$a^3 = 1\angle 360 = 1\angle 0 = 1e^{j2\pi} = 1,\ 1 + a + a^2 = 0,\ \ a + a^2 = -1,$$

$$a^2 - a = -j1.732 = -j\sqrt{3},\ a - a^2 = +j1.732 = +j\sqrt{3}$$

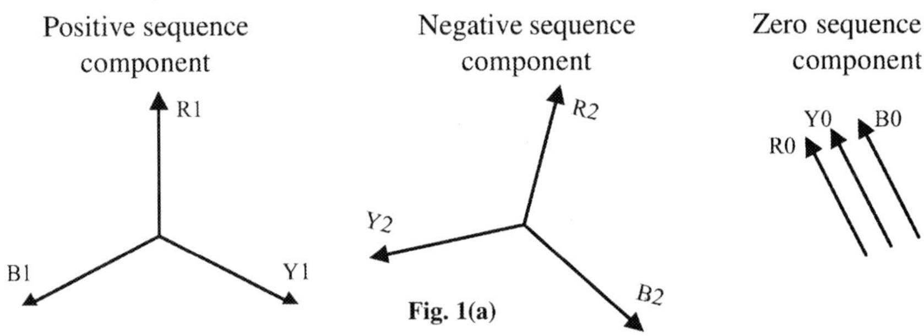

Fig. 1(a)

$V_{R1};\ V_{Y1} = a^2V_{R1};\ V_{B1} = aV_{R1} \qquad V_{R2};\ V_{Y2} = aV_{R2};\ V_{B2} = a^2V_{R2} \qquad V_{R0} = V_{Y0} = V_{B0}$

$V_R = V_{R0} + V_{R1} + V_{R2},$

$V_Y = V_{Y0} + V_{Y1} + V_{Y2} = V_{R0} + a^2V_{R1} + aV_{R2},$

$V_B = V_{B0} + V_{B1} + V_{B2} = V_{R0} + aV_{R1} + a^2V_{R2},$

$$\begin{bmatrix} V_R \\ V_Y \\ V_B \end{bmatrix} = \begin{bmatrix} 1 & 1 & 1 \\ 1 & a^2 & a \\ 1 & a & a^2 \end{bmatrix} \begin{bmatrix} V_{R0} \\ V_{R1} \\ V_{R2} \end{bmatrix}, \quad \begin{bmatrix} V_{R0} \\ V_{R1} \\ V_{R2} \end{bmatrix} = \frac{1}{3} \begin{bmatrix} 1 & 1 & 1 \\ 1 & a & a^2 \\ 1 & a^2 & a \end{bmatrix} \begin{bmatrix} V_R \\ V_Y \\ V_B \end{bmatrix}, \quad \begin{bmatrix} I_{R0} \\ I_{R1} \\ I_{R2} \end{bmatrix} = \frac{1}{3} \begin{bmatrix} 1 & 1 & 1 \\ 1 & a & a^2 \\ 1 & a^2 & a \end{bmatrix} \begin{bmatrix} I_R \\ I_Y \\ I_B \end{bmatrix},$$

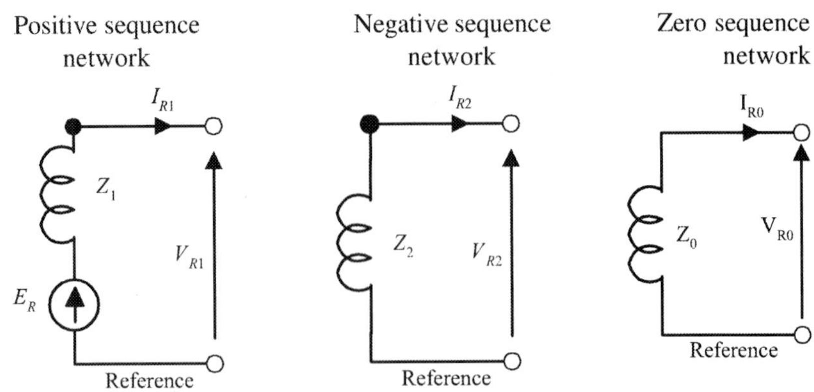

Fig. 1(b)

For line to line fault, as shown in Fig. 1(a), $V_Y = V_B$; $I_R = 0$ and $I_Y = -I_B$. By substituting this, the symmetrical components of voltages are:

$$\begin{bmatrix} V_{R0} \\ V_{R1} \\ V_{R2} \end{bmatrix} = \frac{1}{3}\begin{bmatrix} 1 & 1 & 1 \\ 1 & a & a^2 \\ 1 & a^2 & a \end{bmatrix}\begin{bmatrix} V_R \\ V_Y \\ V_Y \end{bmatrix},$$ so that $V_{R1} = V_{R2}$ and currents are

$$\begin{bmatrix} I_{R0} \\ I_{R1} \\ I_{R2} \end{bmatrix} = \frac{1}{3}\begin{bmatrix} 1 & 1 & 1 \\ 1 & a & a^2 \\ 1 & a^2 & a \end{bmatrix}\begin{bmatrix} 0 \\ -I_B \\ I_B \end{bmatrix},$$ so that $I_{R0} = 0$ and $I_{R1} = -I_{R2}$.

This is shown in Fig. 1(b). Since generator neutral is grounded with a finite impedance Z_0 and $I_{A0} = 0$, $V_{A0} = 0$. From Fig. 1(b),

$$V_{R1} = E_R - I_{R1}Z_1 = I_{R1}Z_2 = V_{R2}$$

$$\therefore\ I_{R1} = \frac{E_R}{Z_1 + Z_2},$$ Now the fault current

$$I_Y = a^2 I_{R1} + aI_{R2} = (a^2 - a)I_{R1} = \frac{(a^2 - a)\,E_R}{Z_1 + Z_2}$$

$$|I_Y| = \frac{\sqrt{3}\,|E_R|}{|Z_1 + Z_2|}$$

Similarly, the open phase voltage V_R is given by

$$\begin{aligned} V_R &= V_{R1} + V_{R2} \\ &= (E_R - I_{R1}Z_1) + (I_{R1}Z_2) \\ &= E_R - I_{R1}(Z_1 - Z_2) \\ &= E_R - \left[\frac{E_R}{Z_1 + Z_2}(Z_1 - Z_2)\right] \\ &= E_R\left[1 - \frac{(Z_1 - Z_2)}{Z_1 + Z_2}\right] \\ &= E_R\left[\frac{2Z_2}{Z_1 + Z_2}\right] \end{aligned}$$

The short circuited phase voltage V_Y is given by

$$\begin{aligned} V_Y &= a^2 V_{R1} + aV_{R2} = (a^2 + a)V_{R1} = -V_{R1} \\ &= -(I_{R1}Z_2) \end{aligned}$$

$$= -\left(\frac{E_R}{Z_1 + Z_2} Z_2\right)$$

$$= -E_R\left[\frac{Z_2}{Z_1 + Z_2}\right]$$

The line voltage $\qquad V_{RY} = V_R - V_Y$

$$= E_R\left[\frac{3Z_2}{Z_1 + Z_2}\right]$$

Now, $\qquad \left|\frac{V_{RY}}{I_Y}\right| = E_R\left[\frac{3Z_2}{Z_1 + Z_2}\right] / \frac{\sqrt{3}E_R}{Z_1 + Z_2}$

$$= \left|\sqrt{3}Z_2\right|$$

$\therefore$ Negative sequence impedance $|Z_2| = \left|\frac{V_{RY}}{\sqrt{3}I_Y}\right|$

CHAPTER 15

To Study the Characteristics of Three-Phase Induction Generator

1. OBJECTIVE

The main objectives of this experiment are:

1. to study the operation of induction motor as a induction generator.
2. to study the effect on power factor and efficiency under induction generator operation.

2. PRE-EXPERIMENTAL QUIZ

1. Write down the equation of torque and power for induction generator.
2. How the positive slip is converted to negative slip?
3. Explain self-excited induction generator.

3. THEORY

3.1 Torque Slip Characteristics

In case of three-phase induction motor, the equation of rotor current is

$$I_2 = \frac{sE_2}{r_2 + jsx_2} \text{ Amp} \tag{1}$$

The equation of torque is

$$T_e = \frac{m}{\omega_s} \bullet \frac{V_{TH}^2}{\left(R_{TH} + \frac{r_2}{s}\right)^2 + \left(X_{TH} + x_2\right)^2} \bullet \frac{r_2}{s} \text{ Nm} \tag{2}$$

The equation of power is

$$P_m = \frac{mV_{TH}^2}{\left(R_{TH} + r_2 + \frac{r_2(1-s)}{s}\right)^2 + \left(X_{TH} + x_2\right)^2} \bullet \frac{r_2(1-s)}{s} \text{ Watts} \tag{3}$$

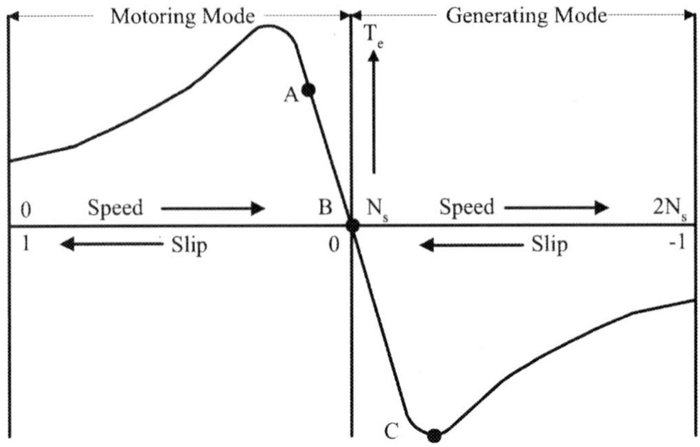

Fig. 15.1. Torque-slip characteristic of a three-phase induction machine showing motoring and generating modes.

Figure 15.1 shows the torque-slip (or torque-speed) characteristics of a three-phase induction motor over its slip range of –1 to 1.

Suppose that a 3-phase induction motor is connected to a constant voltage, frequency supply and is coupled to a prime mover whose speed can be controlled. During the motoring mode, the operating point lies somewhere near point *A*, with positive slip in the range of 2 to 5%. The relative speed between the air gap flux and the rotor conductors is positive, i.e., $(N_s - N_r)$ = +ve.

If the prime mover speed is increased, the operating point begins to travel from *A* towards *B*. At *B*, speed is synchronous N_s, and slip is zero. The relative speed between the air gap flux and the rotor conductors becomes zero, i.e., $(N_s - N_r) = 0$. No rotor emf is induced. The rotor current will decrease to zero, the torque decreases to zero and power also becomes to zero.

If the speed of prime mover is increased further, i.e., speed above synchronous speed, the relative speed between the air gap flux and the rotor conductors becomes negative, i.e., $(N_s - N_r)$ = –ve and slip becomes negative. As a consequence, rotor emf, rotor current, torque and power become negative as per above equations 1, 2 and 3. This implies that for negative slip, electric torque developed by motor is negative, i.e., opposite to the prime mover torque. This implies that at super synchronous speed, rotor does not supply mechanical power to the shaft but absorbs mechanical power from the shaft. Under such condition, machine must act as a generator and delivers its generated power to the supply mains from which it was taking power when working as a motor.

In the above sequence of operation, the induction machine is first operated as motor and runs at sub synchronous speed with positive slip, positive inductance sx_2 and lagging power factor. The motor absorbs both active and reactive power. At super synchronous speed the machine operates as generator with negative slip, negative inductance sx_2 and leading power factor. The generator delivers the active

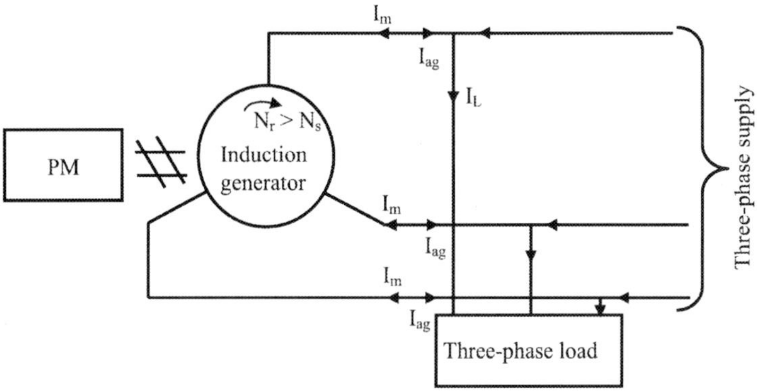

Fig. 15.2. Induction generator connected to three-phase line and load.

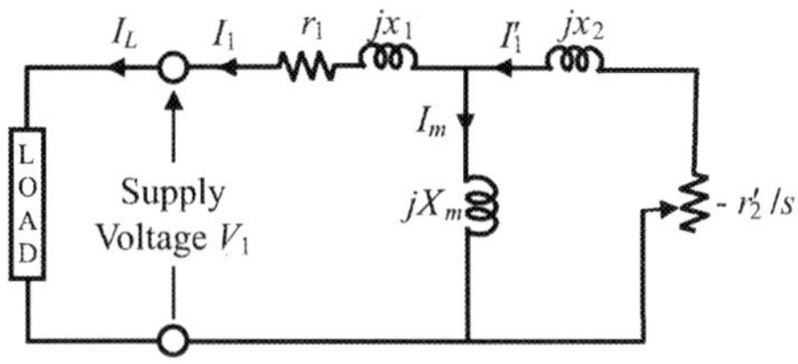

Fig. 15.3. Equivalent circuit diagram for induction generator.

power but absorbs the reactive power, i.e., magnetizing current for the production of flux. The induction generator connected to three-phase line and three-phase load is shown in Fig. 15.2.

At super synchronous speeds, the conditions of an induction generator are similar to those at sub synchronous speeds for induction motor other than negative slip. Therefore the equivalent circuit diagrams and all the equations of induction motor are applicable to induction generator, with negative slip. The torque speed characteristic of induction generator, from slip –1 to 0, is also very similar to that of induction motor, from slip 1 to 0. The point *C* indicates the maximum torque that can be applied to the induction generator by the prime mover. If the prime mover torque is more than the maximum torque, then speed increases, generator goes into unstable zone of the operation and generator current increases. If the protection to the machine is provided by over current relay, then that relay operates and machine stops.

The induction generator is called *asynchronous generator* since the speed during operation is not synchronous.

3.2 Phasor Diagram and Circle Diagram

For prime mover speed above synchronous speed, the rotor is being driven at a speed faster than the synchronously rotating magnetic field. The rotor conductors

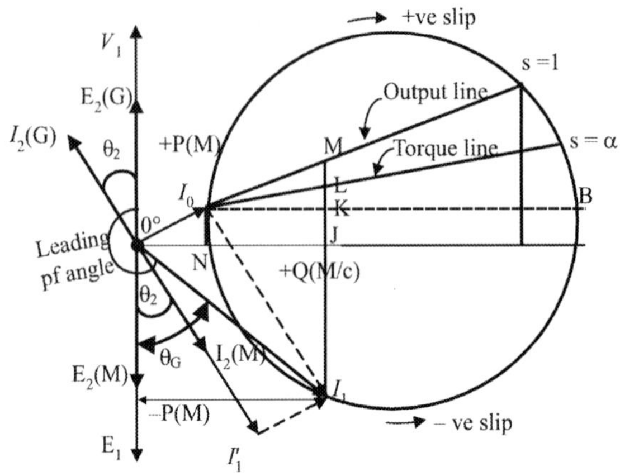

Fig. 15.4. Phasor diagram and circle diagram of induction machine.

are now being cut by the rotating flux in a direction opposite to that during motoring mode. This shows that rotor generated emf E_2, rotor current I_2 and hence its stator component I_1' change their signs as shown in Fig. 15.4. The stator current I_1 is now composed of current I_0 and component I_1'.

In Fig. 15.4 the induction machine is started as motor at point $s = 1$, and comes to the point at I_0. At this point motor is operating on no load, absorbing active and reactive power from the power system. After this point, if the speed is increased, active power decreases to zero and then becomes negative. $E_2 (G)$ and $I_2 (G)$ are the rotor emf and current of the induction generator. The rotor current is reflected on stator side as I_1'. The stator current I_1 is the phasor sum I_1' and I_0.

The upper portion of circle above the dotted line I_0B is applicable for induction motor operation and the lower portion of circle below the dotted line I_0B is applicable for induction generator operation.

For the stator current I_1, JI_1 is the active power which is negative with respect to motor operation, therefore supplying power to the system and OJ is reactive power which is positive with respect to motor operation, therefore power is absorbed by the generator. At point I_l the approximate slip is from –5 to –10%. Input to the generator is MI_1. JK is the constant power loss. KL is the stator copper loss and LM is the rotor copper loss. OJ is the magnetizing current taken from the system which has increased from ON. The power factor cos θ_G is leading for the induction generator.

3.3 Types of Induction Generator

3.3.1 Line Excited Induction Generator

In polyphase induction **motor**, rotating magnetic field is set up by the magnetizing current drawn by the stator from the supply mains. When the speed of motor is made more than its synchronous speed, even then this magnetizing current must

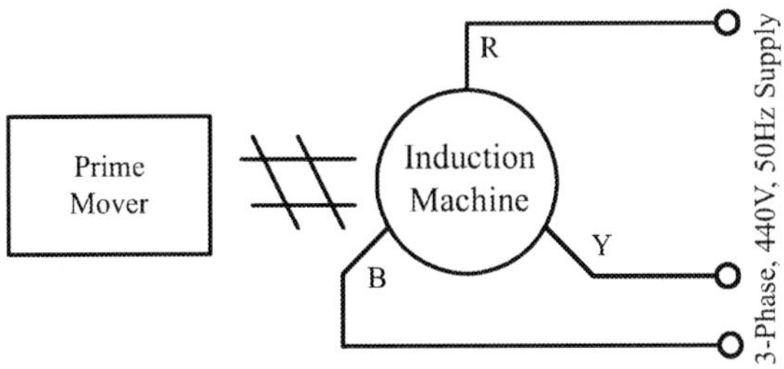

Fig. 15.5. Line excited induction generator.

be delivered to the induction generator by the supply mains so that rotating magnetic field is established. In other words, induction generator **MUST** continue to get its magnetizing current and ***reactive power*** from the supply mains to which it was connected to run as a polyphase induction motor. Thus, three-phase induction generator **CANNOT** work in isolation; it must work in parallel with the bus bar or other synchronous generators which can supply the magnetizing current and reactive power needed by three-phase induction generator. Therefore, it is called *line excited induction generator.*

Under super synchronous speed of motor, if three-phase supply is disconnected, magnetizing current becomes zero, no flux is established, no reactive power is taken from supply and motor may fail to operate as a induction generator. Because of this reason, the double pole double throw switch is connected to the pressure coil and not to the current coil for changing the polarities of the coils in order to measure the power during motor operation as well as generator operation (Fig. 15.9).

3.3.2 Self-Excited or Isolated Induction Generator

If the magnetizing current and reactive power is fed to induction machine by the use of 3-phase capacitor bank, as shown in Fig. 15.6 then at required speed, the induction machine operates as induction generator. This type of generator is called *self excited* or *isolated induction generator.*

This type of generator does not require an existing AC supply system for obtaining its magnetizing reactive power. The capacitor bank provides the reactive power to both, the generator and the load.

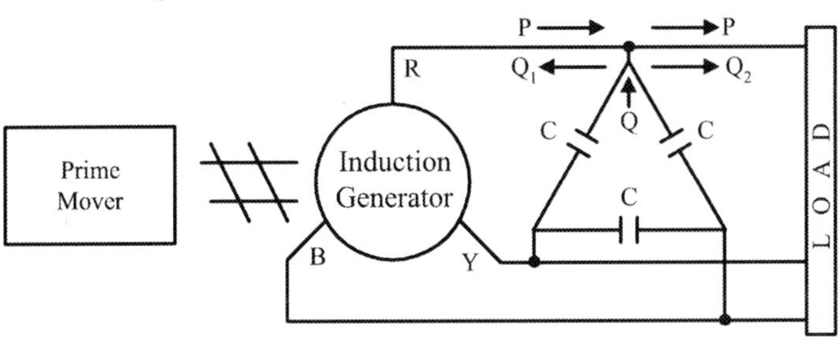

Fig. 15.6. Self-excited or isolated induction generator.

In a self-excited induction generator, stator terminal voltage depends on its magnetization curve, capacitance and speed. When the induction generator is run at the required speed, residual magnetism present in the rotor iron generates a small terminal voltage across the stator terminals. This voltage produces a capacitor current. This current creates a flux which aids the residual flux, thus producing more flux and therefore more voltage across stator terminals. This cumulative process of voltage build up continues till the saturation curve of induction generator intersects the capacitor load line. Note that this voltage build up process is similar to that in DC shunt generator.

If the residual flux is absent in the rotor iron, the induction generator will not build up. This problem can, however, be overcome by running the machine as a motor for some time to create residual magnetism.

In the operation, induction generator is run by prime mover on no load and after building up the voltage load is connected.

The magnetization characteristic and capacitive reactance lines are shown in Fig. 15.7. The intersection gives the no load voltage.

If the generator is run at synchronous speed, with zero slip on no load, then rotor circuit behaves as open and so the circuit model is as drawn in Fig. 15.8. If the stator impedance drop is neglected, then we have $V_1 \approx E$ (induced emf)

$$= I_m X_C \text{ where } X_C = 1/\omega C$$

Thus, the voltage build up depends upon the value of capacitive reactance and magnetizing current. Lower the value of reactance, higher the value of capacitance, greater is the value of no load voltage. In case capacitor line does not intersect the magnetization curve of induction machine, there would be no voltage build up. This will happen when capacitive reactance is very high.

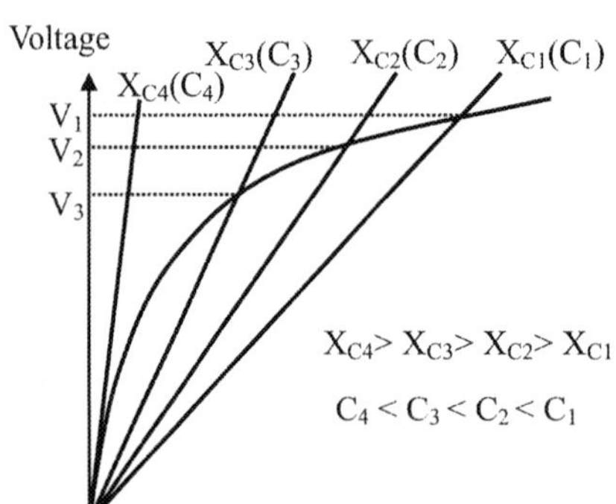

Fig. 15.7. Relationship between capacitance and no load build up voltage.

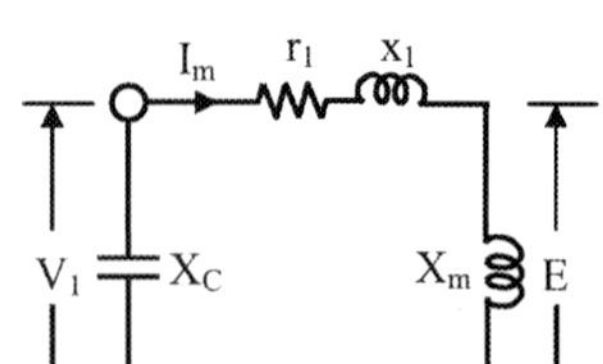

Fig. 15.8. Equivalent circuit diagram for self-excited induction generator.

3.3.3 Advantages and Disadvantages

Induction generators have the following advantages:

1. It is robust in construction.
2. Little auxiliary equipment are required for operation.
3. It can be run in parallel with another generator without hunting at any frequency.
4. Under short circuit, excitation decreases, so short circuit current is limited.

Induction generators have the following disadvantages:

1. More amount of reactive power is required.
2. It has poor efficiency.
3. It can work only at leading power factor.
4. For controlling voltage and frequency, additional units of rectifier and inverter are required.

3.3.4 Applications

1. Used for regenerative braking.
2. Used in mini hydro plants.
3. Used in coastal areas to convert wind energy to electrical energy (wind mills).

4. APPARATUS REQUIRED

1. Induction machine: Rating: (For example, 7.5 HP, 440 V, 1430 rpm)
2. DC Shunt machine: Rating: 5 kW, 230 V, 1600 rpm.
3. Current transformer: One, ratio 20:5.
4. Power factor meter: One, 300 V, 5 Amps
5. AC ammeter: One, 0-20 Amps.
6. AC voltmeter: One, 0-300 Volts.
7. Wattmeter: One, 300 V, 20 A
8. DC ammeter: One 0-25 Amp
9. DC voltmeter: Two, 0-300 Volts and 0-600 Volts, center zero type.
10. Rheostat: Two, 330 Ω, 2 Amps
11. Double pole double throw switch: One.
12. Single pole switch: One
13. Tachometer or speedometer: One

5. CIRCUIT DIAGRAM

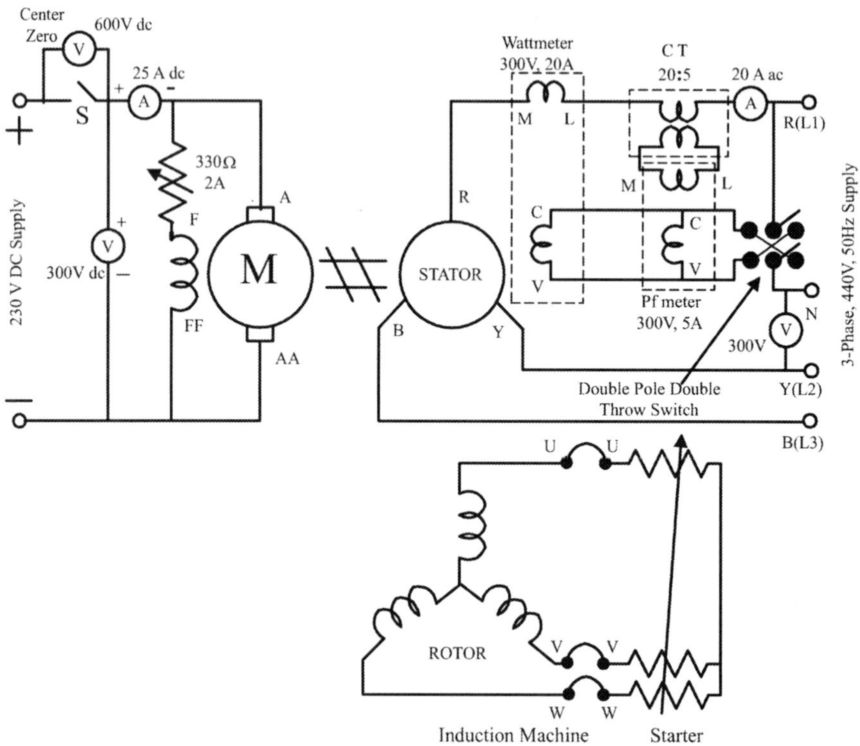

Fig. 15.9. Experimental set-up for studying characteristics of induction generator.

6. PROCEDURE

1. Make the connections as shown in Fig. 15.9 and put *S* off.
2. Set field rheostat of DC shunt machine to maximum resistance point, (OFF), for not to develop armature voltage.
3. Put the starter of induction machine to OFF position, to have maximum resistance in rotor circuit, to produce more starting torque.
4. Switch on AC supply. Press green button of protective or current limiter box to start the **induction machine as a motor (1).**
5. Move the handle of the starter of induction machine towards ON position to decrease the resistance and to increase the speed.
6. Check the DPDT switch position to see that wattmeter records correct input.
7. Decrease the resistance of field rheostat (move the handle from OFF to ON) of DC machine to build up rated voltage of DC machine. **DC machine works as a DC generator (2).**

8. Switch on DC supply and see whether voltmeter across switch *S* reads zero volt or 440 Volts. If it reads 440 Volts, stop the motor and interchange DC supply terminals.
9. If it reads zero Volt, make switch S on.
10. Now, DC generator is connected to DC supply. **Operate DC generator as a DC motor (3).**
11. Increase the speed of DC motor by increasing the resistance of field rheostat (move the handle from ON to OFF, **very very SLOWLY**).
12. As the speed increases, power reading of wattmeter decreases. When the speed is synchronous, wattmeter shows zero output. Above synchronous speed wattmeter shows negative power. **Induction motor operates as an induction generator (4).**
13. In order to read positive power, change the position of DPDT to another side. The power factor changes from lagging to leading for induction generator.
14. At no load or at zero power, note down all the meter readings and the speed.
15. Increase the speed, by moving the handle of field rheostat of DC motor, **very very SLOWLY** from on to off.
16. Note down the speed and readings of all meters.
17. Repeat step no. 15, stepwise upto rated current of induction machine.
18. Repeat step no. 16.

7. OBSERVATION AND CALCULATIONS

DC machine side			Induction machine side				
Speed N	I_{dc}	V_{dc}	I_{ac}	V_{ac}	P_{ac}	PF	η
1500							
1508							
**							
***	Max 25 A						

8. RESULT

The following characteristics are drawn:

1. Speed v_s Output
2. Output v_s I_{ac}
3. Output v_s Power factor
4. Output v_s Efficiency, η

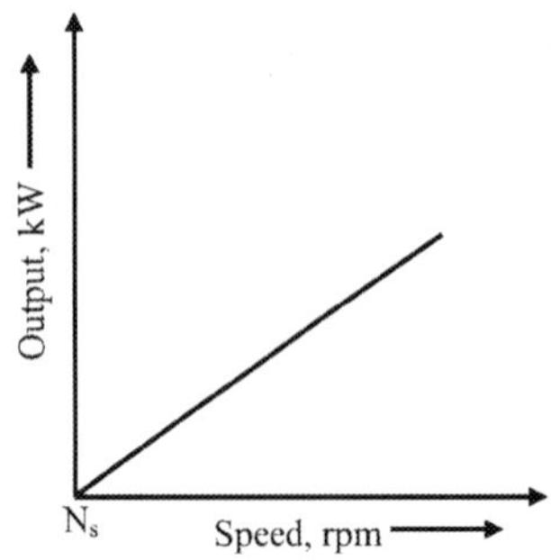

Fig. 5.10. Output vs speed characteristic of IG.

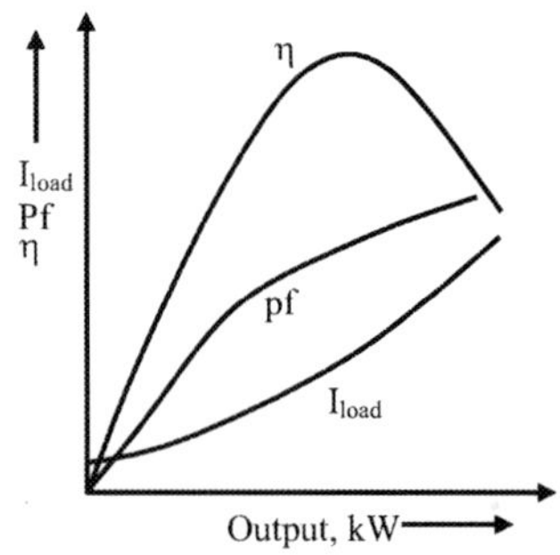

Fig. 5.11. Performance curves of a IG.

9. CONCLUSION

1. As the speed of induction generator increases, output increases upto point *C* of Fig. 15.1.
2. As the output of induction generator increases, stator current I_{ac} increases.
3. As the output increases, power factor is leading and it increases or improves.
4. As the output increases, efficiency increases, becomes maximum and then decreases.

10. POST-EXPERIMENTAL QUIZ

1. The DPDT switch is connected across pressure coil and not in current coil. Why?
2. Why induction generators have leading power factor?
3. Write down the differences between line excited and self-excited induction generator.

CHAPTER 16

To Determine Direct Axis Subtransient (X_d''), Quadrature Axis Subtransient (X_q'') Reactances of Synchronous Machine

1. OBJECTIVE

The main objectives of this experiment are:

1. to determine the direct axis subtransient synchronous reactance and
2. to determine the quadrature axis subtransient synchronous reactance.

2. PRE-EXPERIMENTAL QUIZ

1. Explain constant flux linkage theorem.
2. Explain the symmetrical waveform for armature short circuit current.
3. Define direct axis subtransient synchronous reactance.

3. THEORY

3.1 Short Circuit Analysis

The X_d'' and X_q'' are positive sequence reactances. They appear in the synchronous machine only under short circuit or any other disturbance condition.

The armature winding is a inductive coil, so for simplicity consider a coil of inductance L with negligible resistance r, which is connected suddenly to the sinusoidal source voltage E as shown in Fig. 16.1.

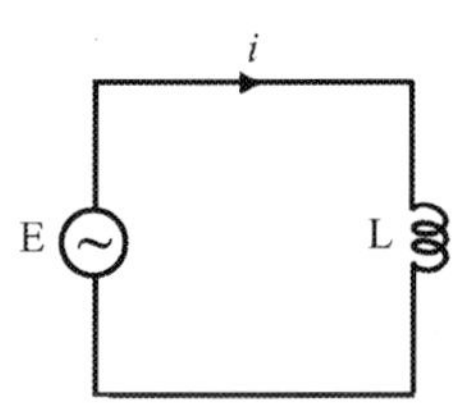

Fig. 16.1. Inductor connected to AC source.

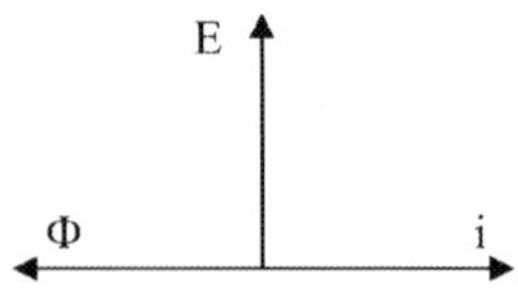

Fig. 16.2. Phasor diagram under short circuit condition.

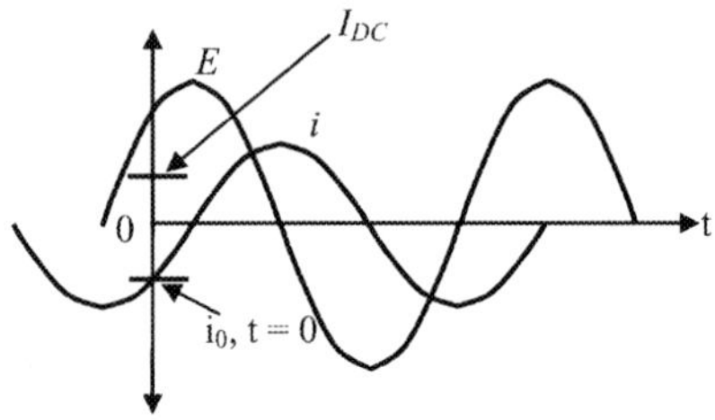

Fig. 16.3. Voltage and current waveform.

The voltage E is induced in the generator due to the flux Φ of field winding. When armature is open circuited, the generator field flux linkages are $\psi_{f0} = L_f I_{f0}$. The flux linkage of inductor L, before the switching on the supply, is zero. Under disturbance condition, the flux linkage of the inductive circuit cannot change ***instantaneously*** due to the opposition property of inductor. Since the flux linkage remains constant at that instant, the theorem related to this is known as ***constant flux linkage theorem.***

Under steady state condition, current lags the voltage E by 90° due to inductor. This is shown in Fig. 16.2 in the form of phasor diagram and in Fig. 16.3 in the form of waveforms.

Suppose that the switch is closed at the instant $t = 0$, as shown in Fig. 16.3. At that instant with respect to voltage E, current i, is having negative value. This current will produce the demagnetizing effect on the flux Φ, since it is in phase opposition to it. The resultant generator field flux will decrease. But as per the constant flux linkage theorem, the flux linkage of inductor L must remain constant (zero in this case) at the instant of switching. This is possible only when there is a immediate flow of current in the circuit, having a magnitude of equal and opposite value, shown as I_{DC} in the Fig. 16.3. This current is called ***DC offset*** current, $I_{DC} = i_0$ at $t = 0$.

If the resistance of the circuit is negligible, then I_{DC} remains constant since time constant is infinity, $\tau = L/r = \alpha$, but if very small value of resistance is considered, then with time constant $\tau = L/r$, the current I_{DC} will decay to zero. The expression for the decaying current is

$$i_{DC} = I_{DC} e^{-t/\tau} \quad (1)$$

The net current finally settles to the steady state current i after I_{DC} becomes equal to zero. The net steady state current is known as the ***symmetrical current***. At the instant of closing the switch the current is symmetrical, if DC offset current is zero.

Now consider the study of short circuit of synchronous generator. The following assumptions are made:

1. The generator is on no load, before short circuit.
2. Armature resistance is negligible.
3. The speed remains constant before and after short circuit. The short circuit transients being electrical in nature are much faster than the electromechanical dynamics. Therefore, major electrical features of short circuit phenomenon are over with respect to speed is considered.
4. All the three phases are short circuited simultaneously.

Now assume that the generator is operating on no load, driven at its rated speed, is excited so that its open circuit voltage is E_f and then its three terminals are suddenly short circuited. As soon as the short circuit is there, the short circuit current is limited by armature leakage reactance similar to transformer.

Just after three-phase sudden short circuit, large current flows and if resistance is neglected, this current will be along d-axis, i.e., lagging by 90° to no load voltage E_f. That's why all d-axis parameters are involved in the short circuit analysis. The demagnetizing mmf F_a must be counterbalanced by a corresponding increase in the field current so that the field flux linkages remain constant at ψ_{f0} as per constant flux linkage theorem. Again this flux cannot change instantaneously as this will take some time to dissipate the stored magnetic energy. This induced component of field current is called **transient** field current i_{ft} which increases the total field current or total excitation. Due to this greater field excitation the short circuit armature current during **transient** period is greater than its value during steady state short circuit.

If the field poles are fitted with **damper bars**, then just after the three-phase sudden short circuit, the demagnetizing mmf F_a is now counterbalanced not only by the induced field current but also by the induced damper bar currents so that field flux linkages remain constant at ψ_{f0}. Now the combined effect of I_{fo}, induced field current i_{ft}, and induced **subtransient** damper bar current i_{dt} again increases the excitation so that under subtransient period due to damper bars, short circuit armature current is greater than its value in transient period.

3.2 Current Waveform

Since the flux linkage of armature winding cannot change instantaneously, the DC offset currents appear in all the three-phases. These currents are dependent to the armature current and flux linkages of each phase at the instant of short circuit. Therefore these are proportional to the cosine of the angle between phase axis and d-axis. For example, if the d-axis is oriented along the R phase axis at the instant of short circuit, then DC offset current in phase R has a certain positive value

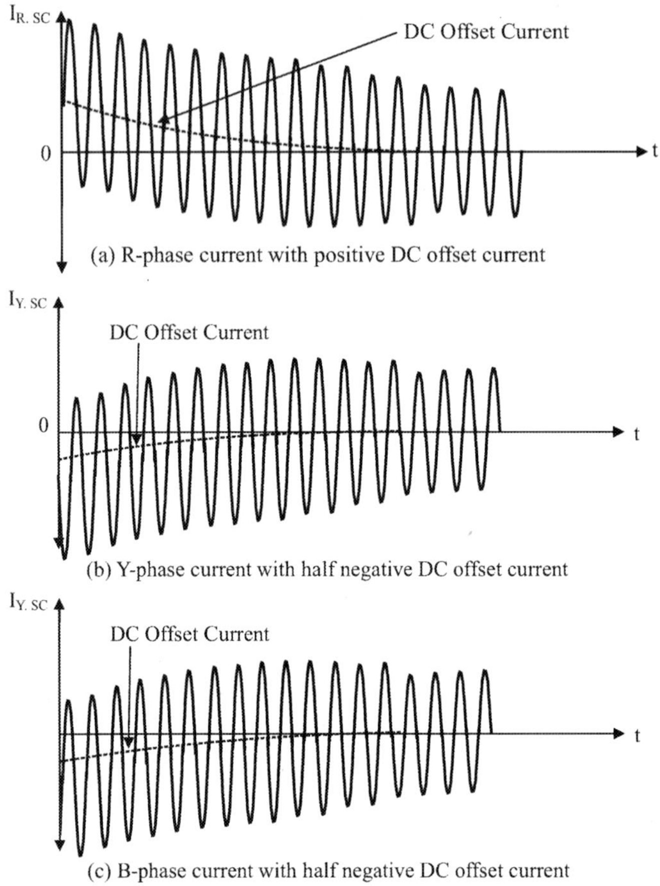

Fig. 16.4. Short circuit current waveforms in three-phases of generator.

($\cos \omega t_0 = \cos \theta_0 = \cos 0 = 1$) and phases Y and B have half negative value ($\cos 120 = -1/2$).

The short circuit current waveforms in the three-phases, R, Y and B are shown in Fig. 16.4 (a), (b) and (c) respectively with the DC offset current in dotted line. If the DC offset current is removed from the short circuit current then the resulting current is called symmetrical short circuit current as shown in Fig. 16.5, which is same for all the three phases but for a phase difference of 120°.

3.3 Symmetrical Short Circuit Current Waveform

As explained earlier the subtransient current is more than transient current and transient current is more than steady state current. The subtransient armature current is due to the steady state field winding current and induced currents in the damper and field windings. Therefore, damper and field windings leakage reactances

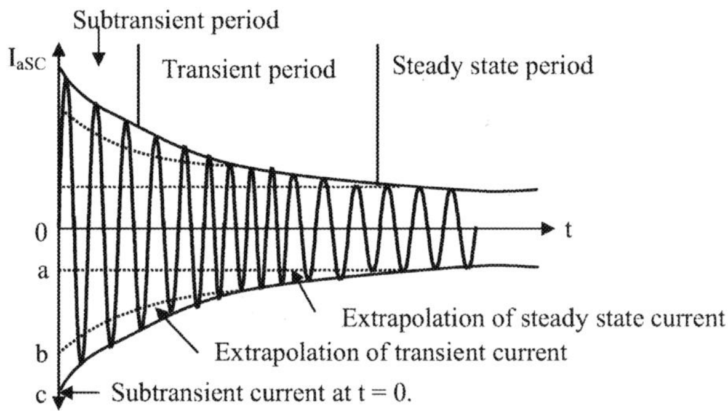

Fig. 16.5. Symmetrical waveform for armature short circuit current of a generator.

are coming in parallel with magnetizing reactance, as far as the operation under subtransient period is considered, similar to induction motor. The equivalent circuit diagram for subtransient condition is shown in Fig. 16.6(a).

The transient armature current is due to the steady state field winding current and induced currents in the field winding. Therefore field winding leakage reactance is coming in parallel with magnetizing reactance, as far as the operation under transient period is considered. The equivalent circuit diagram for transient condition is shown in Fig. 16.6(b).

The steady state current is due to steady state field winding current. The equivalent circuit diagram for steady state condition is shown in Fig. 16.6(c).

The reactances offered by the machine during subtransient period are known as transient reactances, X_d'' – d-axis subtransient reactance and X_q'' – q-axis subtransient reactance. The reactances offered by the machine during transient period are known as transient reactances, X_d' – d-axis transient reactance and X_q' - q-axis transient reactance. The short circuit current is along d-axis, therefore this current is affected by d-axis parameters X_d, X_d' and X_d''.

The induced currents in the damper and field windings decay at the rates determined by their respective time constants. The damper winding consisting of thick bars which has much lower time constant than field winding time constant. The damper winding current decays fastly, so subtransient current decays fastly with damper winding time constant T_d''. The subtransient period lasts for the first few cycles (2 to 3).The induced field current decays with its time constant T_d'. The transient period lasts for relatively many more cycles (8 to 9). Then steady state period starts.

The decaying envelope shows that the d-axis reactance offered by the machine continuously increases as the time progresses and finally settles to steady state value X_d, when the armature reaction demagnetizing action is fully effective. Extrapolation of the subtransient, transient and steady state current envelops identifies the ordinates 0a, 0b and 0c on the current coordinate.

The symmetrical waveform of I_{aSC} shows that the machine presents three different reactances during the short circuit as per the following equations:

$$X_d" = \frac{E_f}{\text{Subtransient current}/\sqrt{2}} = \frac{E_f}{0c/\sqrt{2}} = \frac{E_f}{I_d"}\Omega \tag{2}$$

$$X_d' = \frac{E_f}{\text{Transient current}/\sqrt{2}} = \frac{E_f}{0b/\sqrt{2}} = \frac{E_f}{I_d'}\Omega \tag{3}$$

$$X_d = \frac{E_f}{\text{Steady state current}/\sqrt{2}} = \frac{E_f}{0a/\sqrt{2}} = \frac{E_f}{I_{d.SS}}\Omega \tag{4}$$

It is seen that

$$X_d'' < X_d' < X_d$$

The variation of short circuit current I_{aSC} with time can be expressed as follows:

$$I_{aSC}(t) = (I_d" - I_d')e^{-t/td"} + (I_d' - I_{d.SS})e^{-t/td'} + I_{d.SS} \tag{5}$$

$$I_{aSC}(t) = \left(\frac{E_f}{X_d"} - \frac{E_f}{X_d'}\right)e^{-t/td"} + \left(\frac{E_f}{X_d'} - \frac{E_f}{X_d}\right)e^{-t/td'} + \frac{E_f}{X_d} \tag{6}$$

DC offset armature current establishes flux in the air gap, which produces the fundamental frequency currents in the field and damper windings. This current produces the flux similar to single phase which can be split into two parts: one rotating in the same direction of rotor (suppose clockwise) and other rotating in opposite direction (anticlockwise) at synchronous speed. The clockwise rotating flux is stationary with respect to armature flux and reacts back on the DC offset current. The anticlockwise rotating flux rotates at twice the synchronous speed with respect to the armature field. This induces the second harmonic component (cos 2ωt) of current in the armature winding. With DC offset current and second harmonic current, the resultant short circuit armature current can be written as

$$I_{aSC}(t) = \left[\left(\frac{E_f}{X_d"} - \frac{E_f}{X_d'}\right)e^{-t/td"} + \left(\frac{E_f}{X_d'} - \frac{E_f}{X_d}\right)e^{-t/td'} + \frac{E_f}{X_d}\right]\cos(\omega t + \theta_0)$$

$$+K_1E_f\left(\frac{1}{X_d"} + \frac{1}{X_d"}\right)e^{-t/t_{a1}}\cos(\theta_0) + K_2E_f\left(\frac{1}{X_d"} - \frac{1}{X_d"}\right)e^{-t/t_{a2}}\cos(2\omega t + \theta_0) \tag{7}$$

t_{a1} and t_{a2} are the armature circuit time constants.

3.4 Equivalent Circuit Diagrams

Under short circuit condition, the approximate equivalent circuit diagrams for sub-transient, transient and steady state conditions along d-axis, are shown in Fig. 16.6. (a), (b) and (c) respectively. x_f is the field winding leakage reactance and x_{kd} is the damper winding leakage reactance along d-axis.

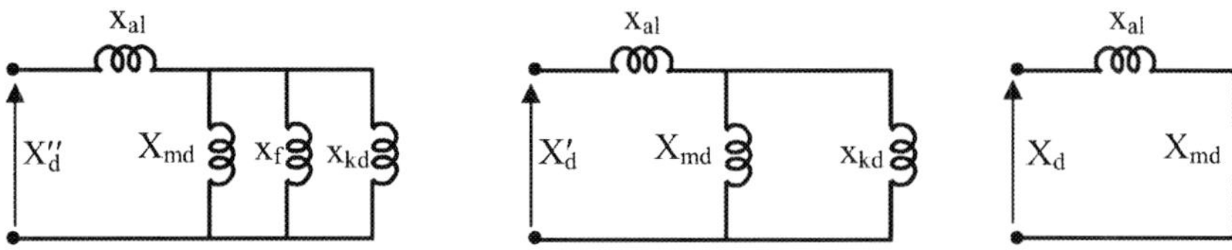

(a) Subtransient condition (b) Transient condition (c) Steady state condition.

Fig. 16.6. Equivalent circuit diagrams for d-axis parameters.

From the equivalent circuit diagram the subtransient reactance X_d'' is given by

$$X_d'' = x_{al} + \frac{1}{\dfrac{1}{X_{md}} + \dfrac{1}{x_f} + \dfrac{1}{x_{kd}}} \tag{8}$$

The transient reactance X_d' is given by

$$X_d' = x_{al} + \frac{1}{\dfrac{1}{X_{md}} + \dfrac{1}{x_f}} = \frac{X_{md} \cdot x_f}{X_{md} + x_f} \tag{9}$$

The steady state reactance X_d is given by

$$X_d = x_{al} + X_{md} \tag{10}$$

If the armature resistance is considered, then the short circuit current will not be along d-axis but slightly lagging to d-axis. In that case, the total short circuit current will be the phasor sum of I_d and I_q. Under short circuit condition, the approximate equivalent circuit diagrams for subtransient, transient and steady state reactances along q-axis can be drawn as shown in Fig. 16.7. (a), (b) and (c) respectively. x_{kq} is the damper winding leakage reactance along q-axis.

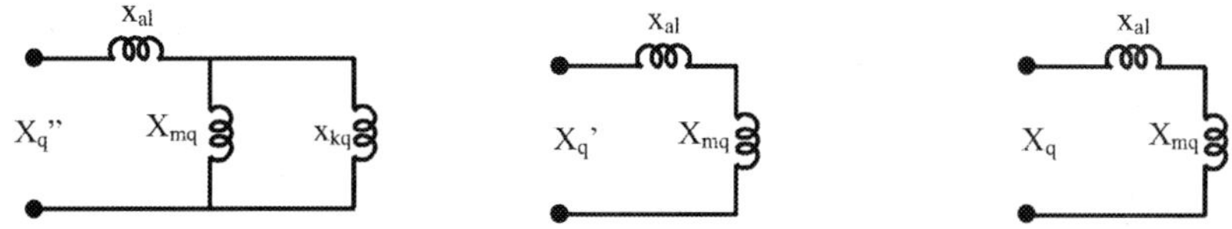

(a) Subtransient condition. (b) Transient condition. (c) Steady state condition.

Fig. 16.7. Equivalent circuit diagrams for q-axis parameters.

Since the field winding axis is the d-axis, there is no effect of field winding along q-axis. Thus the equivalent circuit diagrams for X_q'', X_q' and X_q are similar to those of X_d'', X_d' and X_d respectively but without field circuit parameters.

The structure of damper winding is very similar to squirrel cage rotor of induction motor. There are so many closed circuit and current paths. In order to remove this complexity of so many closed circuits and current paths, the damper winding can be considered to be having only two coils: one coil along d-axis and other coil along q-axis. Therefore, even though the damper winding is housed in the pole shoe, its effect is considered along q-axis and x_{kq} is shown in Fig. 16.7 (a) only in subtransient condition.

From the equivalent circuit diagram the subtransient reactance X_q'' is given by

$$X_q'' = x_{al} + \frac{1}{\dfrac{1}{X_{mq}} + \dfrac{1}{x_{kd}}} = \frac{X_{mq} x_{kd}}{X_{mq} + x_{kd}} \tag{11}$$

The transient reactance X_q' is given by

$$X_q' = x_{al} + X_{mq} \tag{12}$$

The steady state reactance X_q is given by

$$X_d = x_{al} + X_{mq} \tag{13}$$

The transient and subtransient reactances can also be defined as below:

3.5 Direct Axis Transient Reactance, X_d'

The field structure is assumed to have only field winding on it. In order to have field flux linkage zero, the field winding is unexcited and shorted itself. Now the armature winding is suddenly connected to the supply in such a way that the peak of armature rotating mmf wave is along d-axis. The armature flux completes its path through the pole core, stator core and small air gaps. In order to maintain the constant field flux linkage at zero, field current will flow through the shorted field winding, producing the flux which will oppose the armature flux. Therefore the armature flux will not pass through the pole core, but will take the path as shown in Fig. 16.8.

Under this condition, armature flux linkages per armature ampere, is called ***direct axis transient inductance***, L_d' and $X_d' = \omega L_d'$. As compared with the d-axis synchronous reactance, X_d, the flux path of X_d' is having more air gap length, so X_d is more than X_d'.

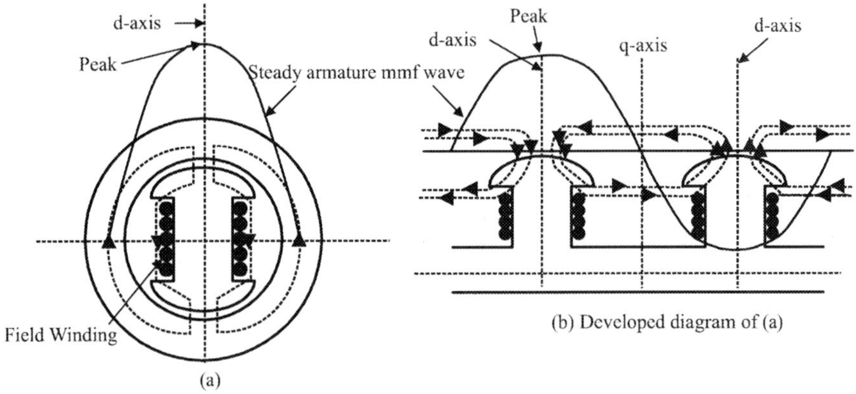

Fig. 16.8. Flux paths for d-axis transient reactance X_d'.

3.6 Quadrature Axis Transient Reactance, X_q'

The conditions of X_d' are applicable to X_q', but in this case the peak of armature rotating mmf wave should be along q-axis. The armature flux completes its path through the pole core, stator core and large air gaps. Since the field winding axis is not along q-axis, there is no effect of armature mmf wave on field winding and the field flux linkage remains at zero (constant). The armature flux will take the path same as that of X_q as shown in Fig. 16.9. Since the air gap is same, $X_q' = X_q$. As compared with the d-axis transient reactance, X_d', the flux path of X_q' is having less air gap length, so X_q' is more than X_d'.

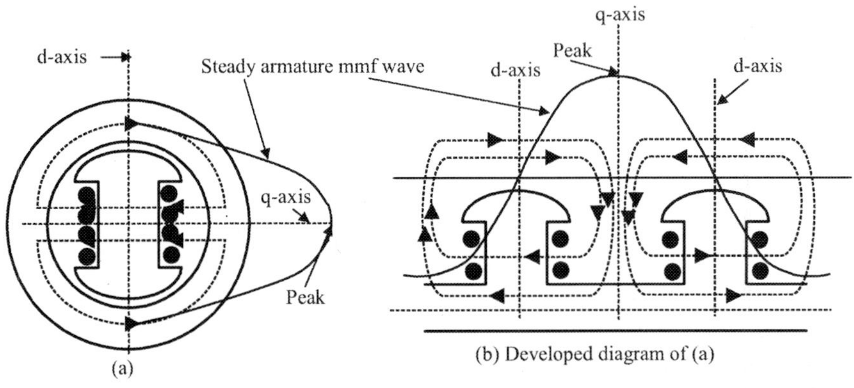

Fig. 16.9. Flux paths for q-axis transient reactance X_q'.

3.7 Direct Axis Subtransient Reactance, X_d''

The salient poles are fitted with damper bars. The cylindrical rotor also provides eddy current paths, producing similar effect as that of damper bars.

The field winding is initially unexcited and closed on itself so that field flux linkages are zero. Now armature currents are suddenly applied in such time phase

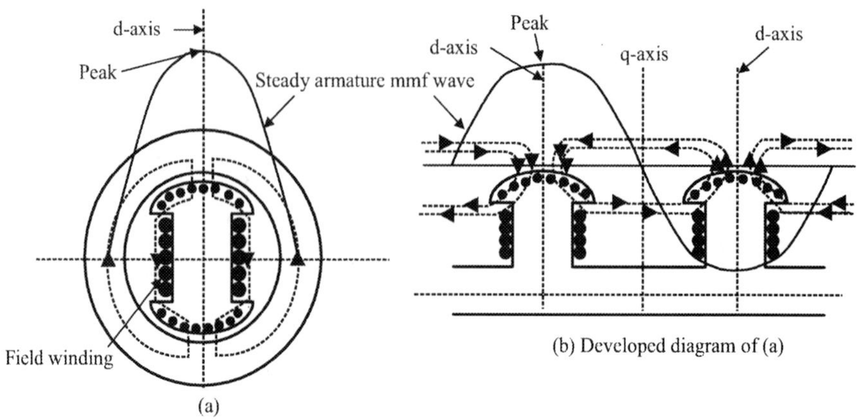

Fig. 16.10. Flux paths for d-axis subtransient reactance X_d''.

that the peak of varying **armature mmf wave is in line with the direct axis.** Since the flux linkages with the rotor circuits are zero, these must remain zero just after the sudden appearance of armature mmf wave. This is in accordance with ***constant flux linkage theorem***. Thus, in order to maintain the flux linkage zero, currents are induced in damper bars, bolts and field winding.

Since the damper bars are near the air gap, the flux lines due to varying armature mmf are forced to follow the leakage flux path mainly in air as shown in Fig. 16.10. Under this condition, armature flux linkages per armature ampere, is called ***direct axis subtransient inductance***, L_d'' and $X_d'' = \omega L_d''$. As compared with the d-axis transient reactance, X_d', the flux path of X_d'' is having more air gap length, so X_d' is more than X_d''.

3.8 Quadrature Axis Subtransient Reactance, X_q''

Here armature currents are applied in such time phase that the peak of varying **armature mmf wave is in line with the quadrature axis**. The damper bars in the q-axis, force the flux lines of the varying armature mmf to follow the leakage paths as shown in Fig. 16.11. The flux linkages with q-axis damper bars must remain zero, before and after the sudden appearance of armature mmf wave. This is in accordance with ***constant flux linkage theorem***. Thus, in order to maintain the flux linkages zero, currents are induced in damper bars.

Under this condition, armature flux linkages per armature ampere, is called ***quadrature axis subtransient inductance***, L_q'' and $X_q'' = \omega L_q''$.

In Fig. 16.11, the armature leakage paths consist of a part of pole face iron and q-axis air space. In Fig. 16.10, the armature leakage paths consist of a part of q-axis air space and whole of the pole face iron. In view of this, for the same armature mmf, the armature flux linkages in Fig. 16.11, are more than in Fig. 16.10. Therefore, X_q'' is slightly higher than X_d''.

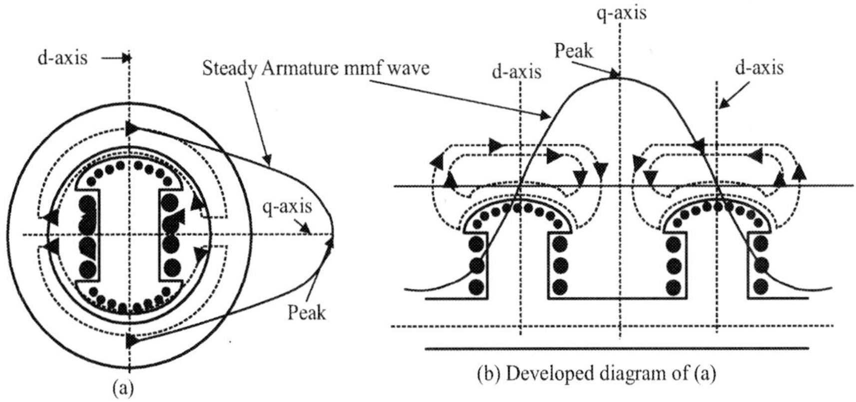

Fig. 16.11. Flux paths for q-axis subtransient reactance X_q''.

Out of all the positive sequence reactances, X_d'' and X_q'' have the lowest magnitude. It is because the damper bars force the armature flux to pass mainly through air (high reluctance path) and as a consequence, armature flux linkages have the minimum values. Therefore X_d'' and X_q'' are the lowest as compared to X_d, X_q, X_d' and X_q'.

The typical values of reactances are: $X_d = 1.2$ pu, $X_q = 1.17$ pu, $X_d' = 0.30$ pu, $X_q' = 1.17$ pu, $X_d'' = 0.15$ pu, $X_q'' = 0.17$ pu, $X_2 = 0.16$ pu, $x_0 = 0.02$ pu.

3.9 Measurement Methods

3.9.1 First Method

Any two phases of synchronous machine are connected in series and a single phase voltage is impressed across them as shown in Fig. 16.12. **The rotor is at standstill**. The impress voltage is adjusted to pass sufficient current in the two series connected armature windings. Now the rotor position is adjusted with hand

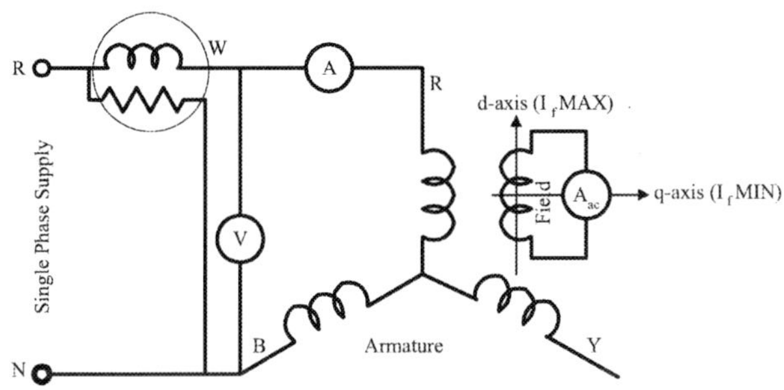

Fig. 16.12. Measurement of X_d'' and X_q''.

to get maximum deflection of the AC ammeter placed in the field winding circuit. Under these conditions, d-axis subtransient impedance

$$Z_d'' = \left[\frac{V}{2I_{aMAX}}\right]_{I_f MAX} \tag{14}$$

The d-axis subtransient reactance,

$$X_d'' = Z_d'' \sin\theta$$

$$= \frac{V}{2I_{aMAX}}\sqrt{1-\left(\frac{P}{VI_{aMAX}}\right)^2} \text{ or } X_d'' = Z_d'', \textit{if } r_a \text{ is neglected} \tag{15}$$

Note that induced current in the short circuited field winding is maximum when field winding axis (d-axis) is along the direction of resultant armature mmf. Since the resultant armature mmf is pulsating in nature the conditions are identical for X_d'' has been defined.

If the rotor shaft is rotated by **HAND** through half a pole pitch or 90° electrical, then the peak of the resultant armature mmf coincides with the q-axis. These conditions are those for which X_q'' has been defined. Since the field winding axis is 90° away from the resultant armature mmf axis, the ammeter in the field winding should record minimum reading. In view of this, for measuring X_q'' the rotor position is adjusted so as to get the minimum value of induced field current. Under these conditions, the instrument readings give, $Z_q'' = \left[\frac{V}{2I_{aMIN}}\right]_{I_f MIN}$. Here I_{aMIN} is armature current corresponding to minimum induced field current.

The q-axis subtransient reactance,

$$X_q'' = Z_q'' \sin\theta$$

$$= \frac{V}{2I_{aMIN}}\sqrt{1-\left(\frac{P}{VI_{aMIN}}\right)^2} \text{ or } X_q'' = Z_q'', \text{ if } r_a \text{ is neglected} \tag{16}$$

3.9.2 Second Method

The X_d'' can also be calculated from symmetrical short circuit armature current waveform. The plot can be obtained from the oscillogram of the short circuit current in any one phase with zero DC offset, as shown in Fig. 16.5. The equation for the calculation of X_d'' is

$$X_d'' = \frac{E_f}{\text{Subtransient current}/\sqrt{2}} = \frac{E_f}{0c/\sqrt{2}} = \frac{E_f}{I_d''}\ \Omega \qquad (17)$$

By knowing X_d'' and negative sequence reactance, X_2, X_q'' can be obtained from the equation

$$X_q'' = 2X_2 - X_d'' \qquad (18)$$

4. APPARATUS REQUIRED

1. Three-phase synchronous machine: Rating: 5 kVA, 440 V, 1500 rpm
2. Dimmerstat: One, 230 V, 30 A.
3. AC Ammeter: Two, 0-30 Amps, 0-1 Amp.
4. AC Voltmeter: One, 0-30 Volts.

5. CIRCUIT DIAGRAM

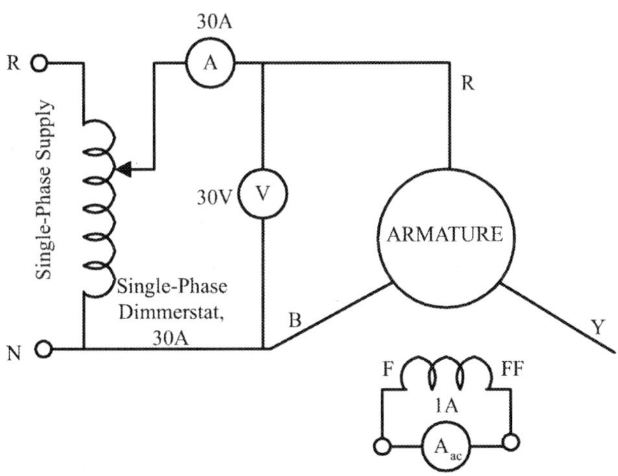

Fig. 16.13. Experimental set-up for determining X_d'' and X_q''.

6. PROCEDURE

1. Make the connections as shown in Fig. 16.13.
2. Set the dimmerstat output to zero and switch on the supply.
3. Gradually increase the dimmerstat output in order to flow armature current less than the rated value.
4. Very very slowly rotate the rotor by HAND and ***simultaneously*** SEE the field ammeter. If this field ammeter shows ***maximum*** current, HOLD the

rotor. Note down voltmeter and ammeter readings. In this case the armature ammeter gives I_{aMAX}.

5. Again slowly rotate the rotor by HAND and ***simultaneously*** SEE the field ammeter. If this field ammeter shows ***minimum*** current, HOLD the rotor. Note down voltmeter and ammeter readings. In this case ammeter gives I_{aMIN}.
6. Repeat steps 4 and 5. Take care that armature current should not exceed its rated value.

7. OBSERVATION AND CALCULATIONS

Observations (for 10A/ph rated)			Calculations			
Voltage V	Armature				Average	Average
	$I_{a\,max}$ $(I_{f\,max})$	$I_{a\,Min}$ $(I_{f\,min})$	$X_d'' = \dfrac{V}{2I_{a\,max}}$	$X_d'' = \dfrac{V}{2I_{a\,min}}$	X_d''	X_q''
	8 A					
	9 A					
	10 A					

8. RESULT

The d-axis and q-axis subtransient synchronous reactances of an alternator are determined and are found to be 1. X_d'' =...... Ω i.e., ...per unit and 2. X_q'' = ... Ω i.e., ...per unit.

9. CONCLUSION

The value of X_q'' is greater than X_d''.

10. POST-EXPERIMENTAL QUIZ

1. Under subtransient period the damper bar effect is considered along q-axis. Why?
2. Can you determine the number of poles of the machine by this method of measurement?
3. The value of X_q'' is more than X_d''. Why?
4. Why subtransient current is more than transient current?
5. All d-axis parameters are involved in short circuit analysis. Why?

CHAPTER 17

Determination of B-H Curve of a Magnetic Material

1. OBJECTIVE

The main objective of this experiment is to study the effects of excitation current on the flux density and permeability of the magnetic material.

2. PRE-EXPERIMENTAL QUIZ

1. Enlist the different magnetic materials.
2. Write down the equation of flux density and magnetizing force.
3. Input current is measured instead of magnetizing force and output voltage is measured instead of flux density. Why?

3. THEORY

3.1 Magnetism and Electromagnetism

In the ancient times people believed that the invisible force of magnetism was purely a magical quality and hence they showed little practical interest. However, with steadily increasing scientific knowledge over the passing centuries, magnetism assumed a larger role. Today magnetism has attained a place of pride in electrical engineering. Without the aid of magnetism, it is impossible to operate devices such as electric generators, motors, transformers, electrical instruments, etc., where electromagnetism is used. Without the use of magnetism, we would be deprived of such valuable assets as the radio, television, telephone, telegraph and the ignition systems of our cars, airplanes, trucks, etc. In fact, electrical engineering is so much dependent on magnetism that without it a few of our modern devices would be possible.

3.2 Magnet and Magnetic Materials

A substance that attracts pieces of iron and steel is called a *magnet* and this property of the material is called *magnetism*. Magnets are found in a natural state in the form of the mineral called *magnetite*. However, natural magnets have no practical value because their magnetism is not strong enough to be utilized in the modern devices. Commercial magnets are made artificially from iron, steel or magnetic alloys.

Those substances which are attracted by magnets are called *magnetic materials*. Iron, steel, nickel, cobalt, and a few special alloys (tungsten, steel, cobalt steel, etc.) are the examples of magnetic materials. Materials which do not exhibit magnetic properties are called *non-magnetic materials*, typical examples being copper, rubber, and wood. Although iron and steel are by far the most common materials for making artificial magnets, yet modern magnets use magnetic alloys, i.e., iron combined with nickel, cobalt, copper or tungsten in order to obtain superior magnetic properties.

3.3 Magnetic Flux, Φ

When we take a bar magnet and dip it into the iron fillings, it is observed that the pieces of iron are attached to the end of the bar magnet. This end of the bar magnet is called the pole of the magnet. A magnet has two poles, one North Pole, *N* and second South Pole, *S*. If a magnet is suspended and brought to rest then the end of magnet pointing to north of the earth is called the South Pole of the magnet and the end of magnet pointing to south of the earth is called North Pole of the magnet.

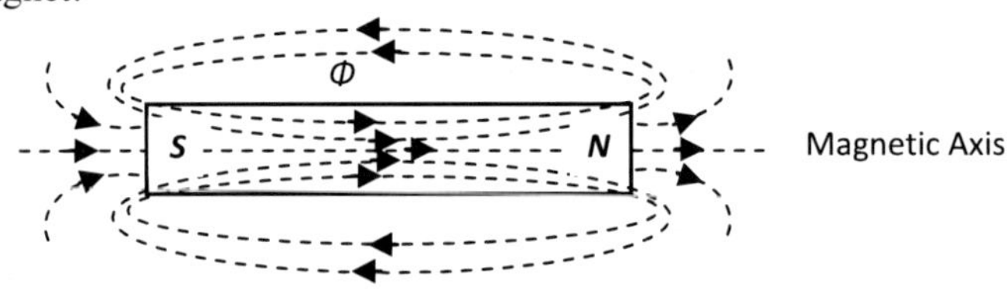

Fig. 17.1. Field around a magnet with lines of force

The line joining these two poles is termed the magnetic axis. The space around the poles is called the magnetic field where poles experience a force. The magnetic field is stronger near the poles and goes on decreasing in strength as we move away from the magnet. The magnetic field around a magnet is represented by imaginary or invisible lines called magnetic lines of force.

The total number of lines of force surrounding a magnet is called total magnetic flux. The magnetic flux is denoted by Greek letter Φ. If 10 magnetic lines come out of the north pole or enter the south pole of a magnet, then magnetic flux $\Phi = 10$ *lines* or *Maxwells*. The SI unit of magnetic flux is Weber and 1Wb = 10^8 lines. The flux is analogous to current in electric circuit.

3.4 Magnetic Flux Density, *B*

The magnetic flux density is the flux per unit area at right angle to the flux lines i.e., magnetic flux density = $B = \Phi/A$ Wb/m^2, *A* is the area in m^2 normal to flux. The other unit of B is tesla. The flux density is a measure of field concentration i.e., amount of flux in each square metre of the field. In fact, it is much more important than the total amount of flux.

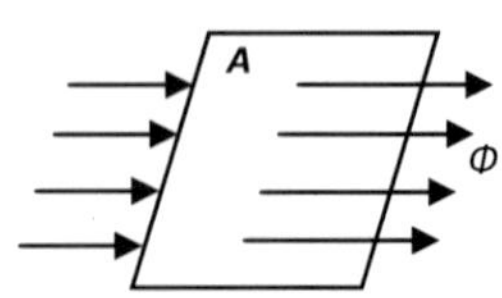

Fig. 17.2. Area *A* normal to the flux Φ

3.5 Magnetomotive Force, mmf

The amount of magnetic flux set up in the core depends on current, I and number of turns, N. If we increase the current or number of turns the amount of magnetic flux also increases. The product NI is called the magnetomotive force, mmf and determines the amount of flux set up in the magnetic circuit.

Therefore, mmf = NI amp-turns

It is a measure of magnetic pressure and is defined as the work done in moving a unit magnetic pole once around the magnetic circuit. It is analogous to emf of an electric circuit and Φ = mmf/S where S is reluctance and analogous to resistance in electric circuit.

3.6 Magnetic Field Strength or Magnetic Field Intensity or Magnetizing Force, *H*

If the magnetic circuit of a magnetic material is homogeneous and of uniform cross-sectional area, the mmf per unit metre length of the magnetic circuit is called the magnetizing force, H.

Therefore, H = mmf / l = NI / l amp per metre

where l is the length of the magnetic flux path in metre.

3.7 Permeability, *μ*

The flux density, B produced in a material is directly proportional to the applied magnetizing force, H. In other words, the greater the magnetizing force, the greater is the flux density and vice versa. Therefore,

$$B \infty H$$

$$B = \mu H$$

Thus, the ratio B/H is always constant and is equal to the absolute permeability of the material, $\mu = \mu_0 \mu_r$, where μ_0 is absolute permeability of vacuum or air, its value is $4\pi\ x\ 10^{-7}$ henry per metre or $H\ /\ m$ and μ_r is relative permeability of the surrounding medium. Its value is 1 for vacuum or air. Thus, when we say that μ_r of soft iron is 8000; it means that for the same magnetizing force the flux density in soft iron will be 8000 times its value in air.

3.8 Reluctance, *S*

The opposition that the magnetic circuit offers to the flux is called reluctance, S. Reluctance in magnetic material corresponds to resistance in an electric circuit. Both of them vary as length / area are depend on the nature of material of the circuit.

In electric circuit, $R = V\ /\ I$ and in magnetic circuit, S = mmf / flux, Φ

Therefore, $$S = \frac{\text{mmf}}{\phi} = \frac{Hl}{Ba} = \frac{l}{\mu_0\, \mu_r\, a}\ AT\ /\ Wb$$

Magnetic materials (iron, steel, etc.) have low reluctance because the value of μ_r is large in their case. On the other hand non-magnetic materials (air, wood, copper, brass, etc.) have high reluctance because they possess least value of μ_r, being 1 in all non-magnetic materials.

3.9 Permeance, Λ (Capital Lambda)

It is the reciprocal of reluctance and is a measure of the case with which flux can pass through the material. Its unit is *Wb / AT*. It is analogous to conductance in electrical circuits. Therefore,

$$\text{Permeance} = \frac{1}{\text{Reluctance}} = \mu_0 \mu_r \frac{a}{l} \text{ Wb / AT}$$

3.10 Types of Magnets

The most usual way of making an artificial magnet is by passing electric current through a wire wound over an iron or steel bar. Depending upon the ability to retain magnetism, artificial magnets are classified as permanent magnets and temporary magnets.

3.10.1 Permanent Magnets

Hardened steel and certain magnetic alloys, e.g., cobalt steel, tungsten steel, etc., when magnetized retain their magnetism indefinitely. Magnets made from such materials are called *permanent magnets*. They have the advantages that no exciting coils are needed and hence no losses. However, their major drawback is that we cannot control the amount of magnetism in the magnet. Permanent magnets are extensively used in electrical instruments, earphones, moving coil, loudspeakers, toys, etc. (Alnico →Alloy – aluminium – nickel – cobalt – iron).

3.10.2 Temporary Magnets (Artificial Magnets)

Iron, soft steel or nickel retain magnetism so long as the magnetizing force is applied. Once the magnetizing force is removed, they will lose nearly all their magnetism. Magnets made from such materials are called *temporary magnets*. The main advantage of such magnets is that, we can control the amount of magnetism by changing the strength of magnetizing agency. Temporary magnets are used in generators, motors, electric bells, relays, dynamic loudspeakers, etc. These are generally horse-shoe or U-shaped because they can retain their magnetism for a long time.

Depending on the value of relative permeability, magnetic materials are classified as diamagnetic materials, paramagnetic materials and ferromagnetic materials.

3.10.2.1 Diamagnetic materials

The materials which are repelled by magnet such as zinc, mercury, lead, sulphur, bismuth, etc., are known as diamagnetic materials. Their relative permeability is slightly less than unity, i.e., 0.9999. They are slightly magnetized when placed in a strong magnetic field and act in the direction opposite to that of applied magnetic field.

3.10.2.2 Paramagnetic materials

The materials which are not strongly attracted by a magnet, such as aluminium, tin, platinum, magnesium, manganese, etc., are known as paramagnetic materials. Their relative permeability is small but slightly in excess of unity (positive). For example, relative permeability of aluminium, air and platinum is 1.00000065, 1.0000031 and 1.00036 respectively. Such materials are slightly magnetized when placed in a strong magnetic field and acts in the direction of the magnetic field.

3.10.2.3 Ferromagnetic materials

The materials which are strongly attracted by a magnet such as iron, steel, nickel, cobalt and some of their alloys are known as ferromagnetic materials. Their **relative permeability** is very high (varying from several hundreds to many thousands). Ferromagnetic materials are of two types.

(a) **Soft magnetic materials:** Those are easily magnetised but do not return their magnetism for any appreciable time after the magnetizing force has been removed.

(b) **Hard magnetic materials:** These materials retain their magnetism with great tenacity, for long time after the magnetising force has been removed and so they are used for making permanent magnets.

Depending on the manufacturing process the magnetic materials are also classified as

1. Cold rolled grain oriented (CRGO) silicon steel and
2. Hot rolled grain oriented (HRGO) silicon steel.

These materials have less eddy current loss. The CRGO has the flux density up to 1.8 Wb / m^2 and HRGO has the flux density up to 1.6 Wb / m^2. When silicon is added to these cores then it is called ferrite core.

3.11 Magnetization Curve

The curve plotted between magnetic flux density *B* and magnetizing force *H* is called "magnetizing curve" as shown in Fig. 17.3 (a). This gives how *B* varies with *H* of any magnetic material. This curve is also called saturation curve or no load curve or simply *B-H* curve. Note the shape of the curve, it is slightly **concave up** for lower flux densities (portion *OC*) and exhibits **straight line character** (portion *CD*) for medium flux densities and reaches knee point. For higher flux densities, the curve **concaves down** (called the knee of the curve). After the knee of the curve, any further increase in *H* does not increase *B*. From this knee, the curve is almost flat and the material is said to be saturated. In terms of domain theory, saturation can be explained as the point at which all magnetic domains are oriented in the direction of applied *H*.

For all non-magnetic materials, μ_r is constant being equal to 1. Therefore, *B-H* curve for a non-magnetic material is a straight line passing through the origin.

The typical variation of μ_r for a magnetic material is shown in Fig. 17.3 (b). Here μ_r is plotted versus flux density rather than *H* because it is onset of the realignment of more and more domains that brings about change in permeability.

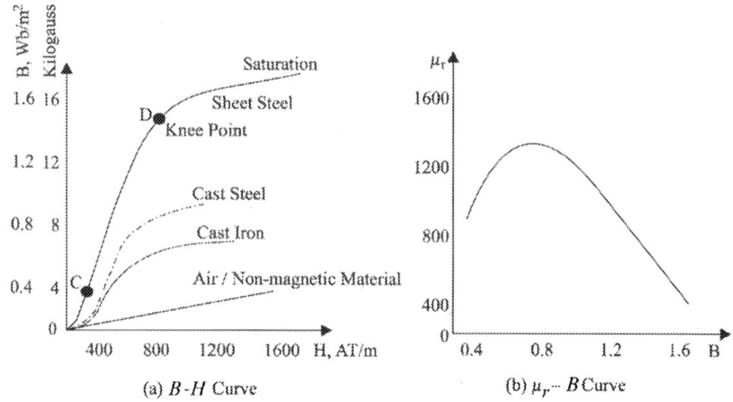

Fig. 17.3. Magnetization curve and variation of relative permeability with flux density

3.12 Magnetic Hysteresis

When a magnetic material is subjected to a cycle of magnetization (i.e., it is magnetized in one direction and then in the other), it is found that flux density *B* in the material lags behind the applied magnetizing force *H*. This phenomenon is called **magnetic hysteresis** (something remains behind).

The term hysteresis derives from the Greek world *hysterein* meaning "lag behind". If a piece of magnetic material is subjected to one cycle of magnetization (compare with one cycle of alternating current or voltage), the resultant ***B-H* curve** is a closed loop and as shown in Fig. 17.3, **abcdefa** is called **hysteresis loop**.

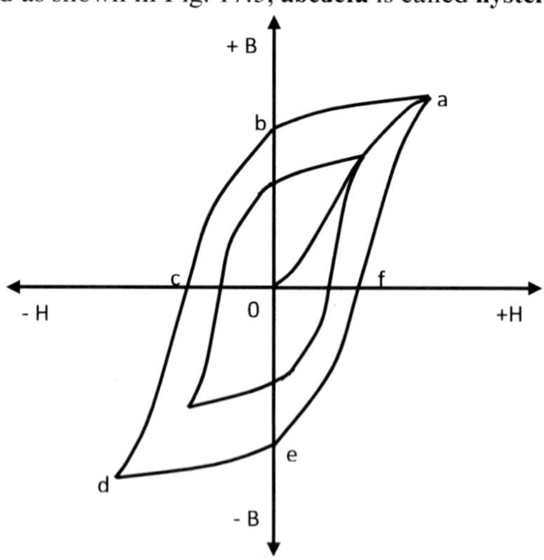

Fig. 17.4. Hysteresis loop

At point 'b', *H* is zero but *B* is not zero in the material but has a finite value +Br, which is called **residual flux density.** This means, after the removal of *H*, the iron piece still retains some magnetism, i.e., +Br. The power of retaining residual magnetism is called **retentivity** of the material or the maximum possible value of the residual flux density is called retentivity.

The value of H required to wipe-out residual magnetism is known as coercive force Hc. The larger the residual magnetism, the larger must be the coercive force. The maximum value of the **coercive force** is called **coercivity.**

When a magnetic material is subjected to a cycle of magnetization, an **energy loss** takes place due to molecular friction in the material. This loss is in the form of heat and is called **hysteresis loss.** It is so called because it results due to the hysteresis effect in the magnetic material.

The area of hysteresis loop represents the **energy loss / m³** during one cycle of magnetization (energy density). The area of loop is the product of B and H. The wider loop characteristic is used for permanent magnets and the narrow loop characteristic is used for electrical machines.

3.13 Measurement Method

For the measurement of B and H, a transformer is used. There are two windings wound on a laminated core of a magnetic material. The winding, which is connected to an alternating source voltage or the winding which receives energy from the source of AC supply is called *primary winding.* The winding which is connected to the load and delivers energy is called *secondary winding.*

An alternating source voltage V is given primary winding, an alternating current I starts flowing through N_1 primary turns. The alternating mmf NI sets up an alternating flux Φ. Thus, H is directly proportional to I if N and l are constant.

Let the sinusoidal flux be given by

$$\Phi = \Phi_{max} \, \text{Sin}\omega t$$

where Φ_{max} is the maximum value of core flux.

As per Faraday's law, the instantaneous emf (e_1) is induced in primary winding and emf (e_2) is induced in secondary winding. The expression for emf, e_2 is given by

$$e_2 = N_1 \frac{d\Phi}{dt}, \quad e_2 \propto \Phi, \; e_2 \propto B$$

Hence, the nature of V-I curve will be same as that of B-H curve.

4. APPARATUS REQUIRED

1. Single Phase Variac / Dimmerstat / Autotransformer: 230V, 2-4 Amp.
2. Transformers: 1. Iron Core 2. Ferrite Core: 230V, 1-2 Amp.
3. AC Ammeter (Moving Iron Type: used for AC & DC): One, 1-2 Amp.
4. AC Voltmeter (Moving Iron Type: used for AC & DC): One, 300 V.

5. CIRCUIT DIAGRAM

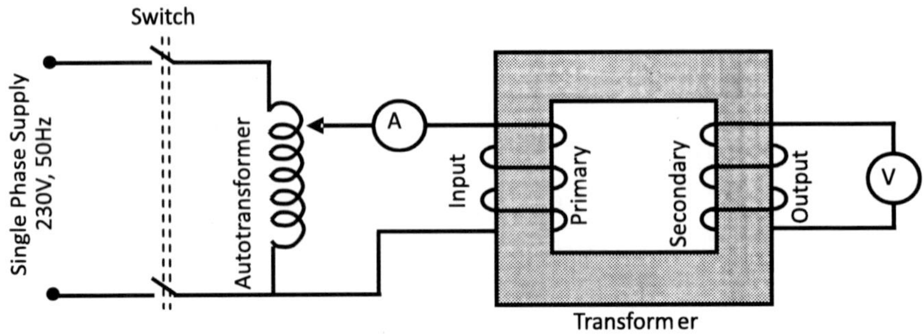

Fig. 17.5. Experimental setup for determination of *B-H* curve of a magnetic material

6. PROCEDURE

1. Make the circuit connections as shown in Fig. 17.5.
2. Set the dimmerstat output to zero and switch on the single phase supply.
3. Gradually increase the dimmerstat output voltage in order to flow the current less than the rated value in primary winding.
4. Note down the current in primary winding and induced voltage in the secondary winding.
5. Increase the output voltage of dimmerstat to a suitable value.
6. Repeat step 4.
7. Take readings up to rated current of primary winding of transformer.
8. Repeat all the above steps if another transformer with another magnetic material is available.

7. OBSERVATIONS

1. Transformer (with iron core) 2. Transformer (with ferrite core)

Sr. No.	Current in primary winding Amp	Voltage in secondary winding Volt
1.		
2.		
..		

8. RESULTS

From the values of current, *I* and voltage, *V*, draw $B(V)$-$H(I)$ curve.

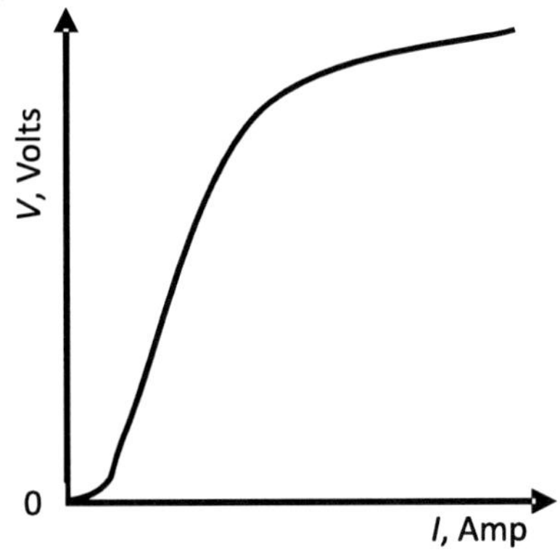

Fig. 17.6. *B-H* Curve

9. CONCLUSION

1. As the current in primary winding increases, voltage in secondary winding increases up to knee point of *B-H* curve.
2. Up to knee point, *B-H* curve is straight line characteristic.
3. After knee point, as the current increases the voltage becomes constant.
4. The relative permeability initially increases up to knee point and then decreases in saturation zone.
5. The decrease in relative permeability in saturation is not good for the operation of transformer.

10. POST-EXPERIMENTAL QUIZ

1. Mention knee point and draw air gap line in *B-H* curve.
2. What is the effect of saturation on permeability of a magnetic material?
3. What are the bad effects of saturation on the performance of a transformer?

CHAPTER 18

To Study and Verify Kirchhoff's Laws Applied to Direct Current Circuits

1. OBJECTIVE

The main objective of this experiment is to study the Kirchhoff's laws applied to direct current circuit and to verify it.

2. PRE-EXPERIMENTAL QUIZ

1. State KCL and KVL.
2. What is node?
3. What is the difference between mesh and loop?

3. THEORY

By the middle of the nineteenth century a considerable amount of experience with electric circuits had been gained by the leading men of science of the time. However, it was Gustav Robert Kirchhoff (1824–1887), an eminent German physicist, who published the first systematic formulation of the principles governing the behaviour of electric circuits. He advanced no new experimental facts or concepts but merely restated familiar principles. His work was embodied in two laws – a current and a voltage law–which together are known as Kirchhoff's laws. It is upon these laws that **electric circuit** theory is based.

3.1 DC Circuit

It is well known that electric current flows in a closed path. The closed path followed by electric current is called **electric circuit**. The essential parts of an electric circuit are:

1. The source of power (e.g., battery, generator, etc.)
2. The conductor used to carry current
3. The load (e.g., lamp, heater, motor)

The device which uses the electrical energy is called **load**. The source supplies electrical energy to the load which converts it into heat or other forms of energy. Thus, conversion of electrical energy into other form of energy is possible only with suitable circuits. For instance, conversion of electric energy into mechanical energy is achieved by dividing a suitable motor circuit. The heater converts electrical energy supplied to it into heat.

The closed path followed by direct current (DC), is called **DC circuit**. A DC circuit essentially consists of a source of DC power (e.g., battery, DC generator), the conductors used to carry current and the load connected to a battery through conducting wires. The direct current starts from the positive terminal of the battery and comes back to the starting point via the load. The load for a DC circuit is usually a resistance. Other passive elements, i.e., inductance and capacitance are relevant only in AC circuits.

In a DC circuit, loads may be connected in series or parallel or series parallel. Accordingly DC circuits can be classified as:

1. Series circuits
2. Parallel circuits
3. Series-parallel circuits

3.2 Kirchhoff's First Law or Current Law or Point Law

1. The sum of currents entering or leaving a junction point at any instant of time is equal to zero.
2. The algebraic sum of currents flowing towards a junction in an electric circuit is zero
3. The sum of the currents flowing towards any junction in an electric circuit is equal to the sum of currents flowing away from the junction.

A junction point is that place in a circuit where two or more circuit elements are joined together. It is often called an independent node.

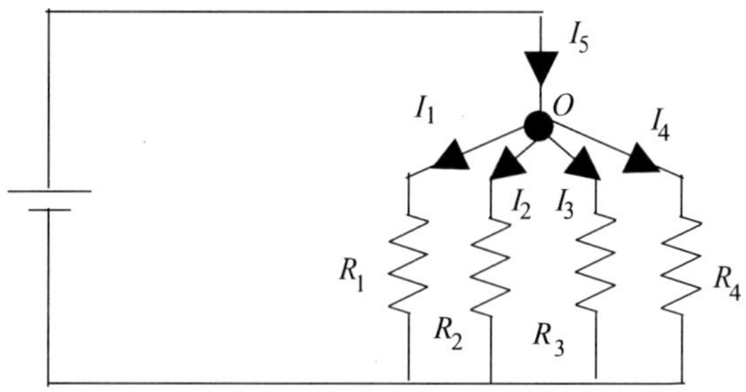

Fig. 18.1. Kirchhoff's current law

In Fig.18.1, O corresponds to the junction point or node. The current law may be expressed mathematically as

$$\sum_{i=0}^{k} I_i = 0$$

where k denotes the number of circuit elements connected to the node in circuit and Σ is the Greek symbol to indicate summation. In Fig 18.1, there are five elements connected to node O – four passive elements (resistors) and one active element (battery).

According to KCL,

$$I_5 - I_1 - I_2 - I_3 - I_4 = 0$$

The minus signs are used because these currents are defined as leaving rather than entering the node.

Therefore,

$$I_5 = I_1 + I_2 + I_3 + I_4$$

Incoming current = Outgoing current

In countless experiments performed over the past century Kirchhoff's current law has never been found to be invalid. This is understandable when it is realized that the current law is nothing more than a restatement of the **principle of conservation** of charge. This principle states that the number of electrons passing per second must be the same for all points in the circuit. Accordingly, if the summation of all the currents at a node were not to add up to zero, there would have to occur an **accumulation of charge** at the node. For the sake of argument suppose the sum of currents at node *O* were not zero but rather one ampere. Then charge would accumulate at the rate of 1 coulomb for each second. We have already learnt, however, that by Coulomb's law a charge accumulation of thin kind would produce explosive forces. But experiment shows no evidence of thin layer whatsoever. Any accumulation of charge at the node means accumulation of mass. Once again experiment fails to reveal that for every charged particle that enters the node there is another charged particle that leaves.

3.3 Kirchhoff's Second Law or Voltage Law

At any instant, the algebraic sum of all the electromotive force (emf) and the voltage drops in a closed circuit is zero.

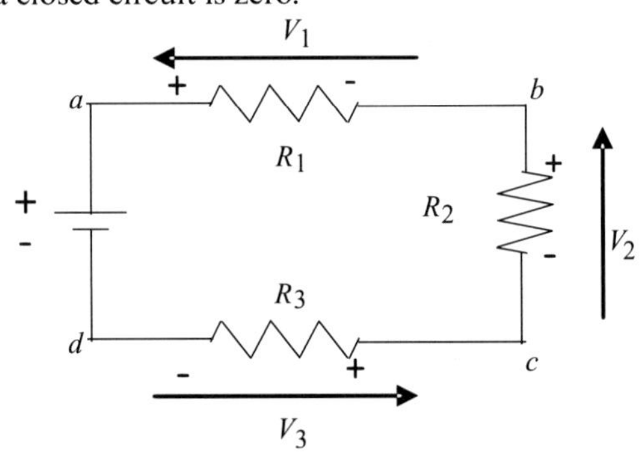

Fig. 18.2. Kirchhoff's voltage law

Essentially, this law is a restatement of the law of **conservation of energy**. To understand this law refer to Fig. 18.2, which depicts 12V battery supply energy to three series connected resistors representing a total resistance of 12Ω. By Ohm's law there exists a current of 1A, which means that charge flows throughout the circuit at the rate of 1 coulomb per second. This current is denoted by *I*.

The positive direction of current flow is a-b-c-d-a. The electrons of course, are moving in the reverse sense. The constant current which flows in this circuit means that the average rate of movement of charge is constant. However, as electrons traverse the circuit elements R_1, R_2 and R_3 they collide with atoms in the space lattice configurations of these elements, thereby dissipating energy in terms of heat. Since this energy must be the energy which permits us to write

$$W = W_1 + W_2 + W_3$$

where W denotes the energy supplied by the battery in a specified time interval, W_1 denotes the energy dissipated in R_1 in the same interval of time, and similarly for W_2 and W_3. Furthermore, if we denote by Q, the total charge which moves throughout the circuit during this same interval of time, it follows that the work per unit charge may be written as

$$\frac{W}{Q} = \frac{W_1}{Q} + \frac{W_2}{Q} + \frac{W_3}{Q}$$

But work per unit charge in voltage

$$E = V_1 + V_2 + V_3$$

or

$$E - V_1 - V_2 - V_3 = 0$$

A generalized formulation of Kirchhoff's voltage law written as

$$\sum_{i=0}^{k} V_i = 0$$

where V_i represents the voltage drop of the ith element in any given closed circuit which is assumed to have k elements.

While applying KVL, a closed circuit or mesh is considered. Therefore, it is important to assign proper signs to emfs and voltage drops in the closed circuit. The following sign convention may be followed:

A rise in the potential should be considered positive while a fall in potential should be considered negative.

The reverse convention is equally valid, i.e., rise in potential may be considered negative and fall in potential as positive.

In a closed circuit, if we move from negative to positive terminal, then there is rise in the potential and if we move from positive to negative then there is fall in potential. In a resistor, entering current terminal is positive and leaving current terminal is negative, because current flows from higher potential to lower potential.

3.4 Application

First of all the current distribution in various branches of the circuit is made with directions of their flow complying with first Kirchhoff's law. Then Kirchhoff's second law is applied to each mesh (one by one) separately and algebraic equations are obtained by equating the algebraic sum of emfs acting in a mesh equal to the algebraic sum of respective drops in the same mesh. By solving the equations so obtained unknown quantities can be determined while applying Kirchhoff's second law, the question of algebraic signs may be troublesome and is a frequent source of error. If, however, the following rules are kept in mind, no difficulty should occur.

1. The resistive drops in a mesh due to current flowing in clockwise direction must be taken as positive.
2. The resistive drops in a mesh due to current flowing in counter-clockwise direction must be taken as negative.
3. Similarly, the battery emf and the battery emf causing current to flow in counterclockwise direction in a mesh must be taken negative emf.

For example for the current shown in Fig.18.3, let the current distribution be made as shown which satisfies Kirchhoff's first law fully.

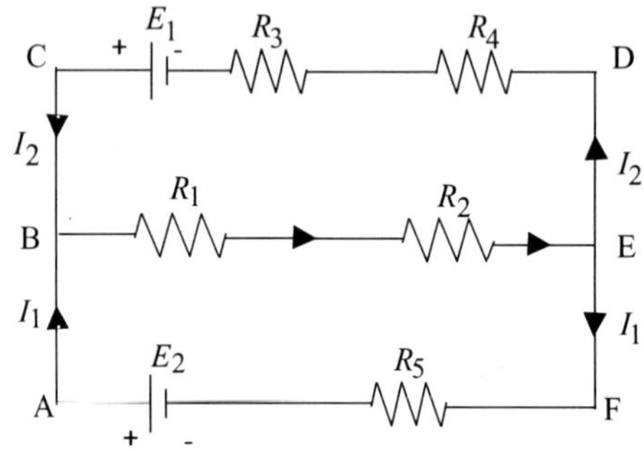

Fig. 18.3. Circuit diagram for application of Kirchhoff's laws

Taking the first mesh ABEFA for the application of Kirchhoff's second law, we see that there is only one emf acting in the mesh (E_2) and since it tries to send current in clockwise direction so E_2 be taken as positive, similarly all the resistive drops, i.e., $R_1(I_1+I_2)$, $R_2(I_1+I_2)$ and R_5I_1 are clockwise, so these must be taken as positive.

According to Kirchhoff's second law in mesh ABEFA

$$E_2 = R_1\,(I_1 + I_2) + R_2\,(I_1 + I_2) + R_5 I_1$$

In mesh BCDEB, there is only one emf acting in the mesh (E_1) and since it tries to send current in counterclockwise direction through the mesh under consideration so it may be taken as negative. Since all the resistive drops R_3I_2, R_4I_2, $R_2(I_1+I_2)$, $R_1(I_1+I_2)$ are counterclockwise, so these may be taken as negative.

Hence, according to Kirchhoff's second law in mesh BCDEB we get,

$$-E_1 = -R_3I_2 - R_4I_2 - R_2(I_1 + I_2) - R_1(I_1 + I_2)$$

$$E_1 = R_3I_2 + R_4I_2 + R_2(I_1 + I_2) + R_1(I_1 + I_2)$$

$$E_2 - E_1 = -R_3I_2 - R_4I_2 + R_5I_1$$

$$E_2 - E_1 = R_5I_1 - (R_3 + R_4)I_2$$

Thus, depending on the number of meshes, voltage equations are written. If the values of voltages and resistances are known, then by solving simultaneous voltage equations, currents in various branches or resistors are known.

4. APPARATUS REQUIRED

1. DC ammeter (PMMC or MI type): three, (0-1)A
2. Variable resistors : three, 10 ohms

3. Regulated power supply : two
4. Digital multimeter: one

5. CIRCUIT DIAGRAM

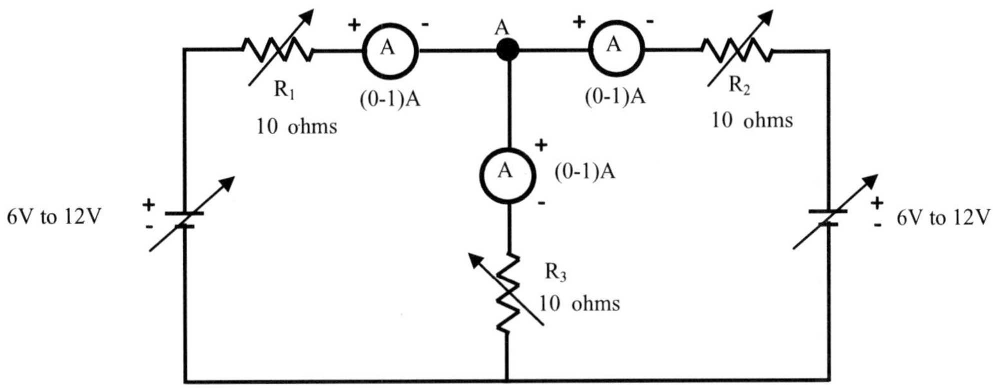

Fig. 18.4. Experimental set-up for verification of Kirchhoff's laws

6. PROCEDURE

1. Make connections as per the circuit diagram.
2. After the connections are checked, make on both the supplies. Adjust the voltages to 6V to 10V.
3. Note down the voltages across the resistances R_1, R_2, R_3, as V_{R1}, V_{R2}, V_{R3} and measure the voltage of the supplies as E_1 and E_2 and enter in the observation table.
4. Note down the current through R_1, R_2 and R_3.
5. Change the value of the resistances till you notice appreciable change in ammeter readings.
6. Repeat steps 3, 4, 5 to complete 3 sets of readings.
   The readings taken in steps 3 and 4 comprise one set of reading.

7. OBSERVATIONS

Sr. No.	I_1 Amp	I_2 Amp	I_3 Amp	V_{R1} Volt	V_{R2} Volt	V_{R3} Volt	E_1 Volt	E_2 Volt
1.								
2.								
3.								
4.								

8. CALCULATIONS

Verification of KCL	At junction A	$: I_1 + I_2 + I_3 = 0$
Verification of KVL	Loop 1	$: E_1 = V_{R1} + V_{R3}$
	Loop 2	$: E_2 = V_{R2} + V_{R3}$

9. RESULTS

KCL and KVL are verified.

10. POST-EXPERIMENTAL QUIZ

1. How will you decide the direction of current and polarity of voltage drop in a resistor?
2. Can Kirchhoff's laws be applied to AC circuits?
3. What are the factors ignored in the calculations?

CHAPTER 19

To Study R-L-C Series Circuit

1. OBJECTIVE

The main objective of this experiment is to study the effect of resistive, inductive and capacitive circuit on current phasor and power factor.

2. PRE-EXPERIMENTAL QUIZ

1. What do you mean by inductive and capacitive reactance?
2. Define power factor?
3. Draw the waveforms and phasor diagrams for voltage and current for resistor, inductor and capacitor?

3. THEORY

When the flow of current in circuit is in one direction, then the current is called as Direct Current, DC. But if the current flows in both the directions at regular intervals, then the current is called as Alternating Current AC and the circuit is called as AC Circuit. In AC system, the voltage changes its polarity at regular interval of time and the resulting current changes direction accordingly. The waves of alternating voltage and current are sinusoidal.

3.1 Phase

The waves of alternating voltage and current are continuous. They do not stop after one cycle but continue appearing as long as the AC generator is operating. The time starts from the instant when the value of voltage or current is zero. In sinusoidal voltage, the maximum positive value is $+V_m$ and occurs at 90° or $\pi/2$ radians or T/4 seconds. Therefore, it is said that the ***phase*** of maximum positive value $+V_m$ is 90° or $\pi/2$ radians or T/4 seconds. For 50 Hz supply it is 5 milliseconds. Similarly, the ***phase*** of maximum negative value $-V_m$ is 270° or $3\pi/2$ radians or 3T/4 seconds or 15 milliseconds.

Hence, the ***phase*** of an alternating quantity is a fractional part of time period or cycle through which the wave has advanced from selected zero position of reference. The following points may be noted:

1. The ***phase*** of an alternating quantity depends on the instant from which time is measured, i.e., zero.
2. An alternating quantity is completely known if we know its maximum value, frequency and phase.

3.2 Phase Difference

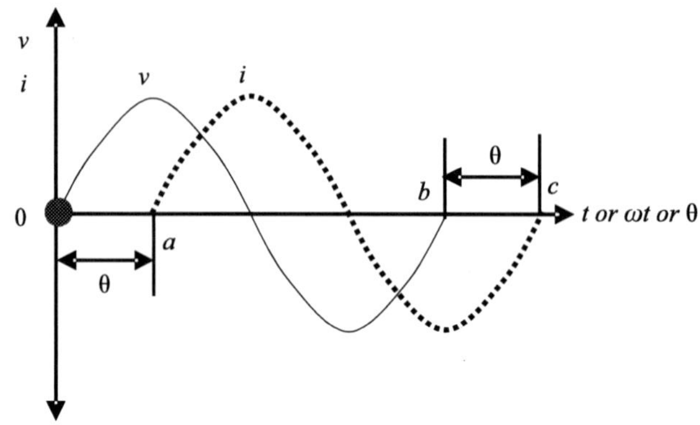

Fig.19.1. Phase difference

When the alternating voltage is applied to a circuit, an alternating current of same frequency flows through the circuit. But the current value may or may not be equal to zero when the value of voltage is zero. In other words, we say that voltage and current have a phase difference.

Hence, when two alternating quantities of the same frequency have different zero points, they are said to have ***phase difference***.

The angle between zero points is the angle of ***phase difference*** θ. It is generally measured in degrees or radians. In the positive half cycle, the wave which passed through its zero point earlier is said to be ***leading*** while the other is said to be ***lagging***.

In Fig. 19.1, voltage v is passing through its zero point, 0 and rising in the positive direction. Similarly, current i passes through its zero, a, and rising in the positive direction. Therefore, phase difference between voltage and current is $0a = \theta$ or $bc = \theta$. The current is lagging to the voltage by an angle θ. The equations of voltage and current are:

$$v = V_m \sin\omega t \text{ and } i = I_m \sin(\omega t - \theta)$$

The following points may be noted:

1. The minus sign before θ is used because current lags behind the voltage. A plus sign would be used if the current is leading to the voltage.
2. Note that ωt is variable and θ is constant. This means phase difference remains constant.
3. Usually ωt is expressed in radians and θ in degrees. In engineering calculations, the phase angle θ is important and not the total angular displacement.

3.3 Phasor Representation of Sinusoidal Quantities

Often in determining the sinusoidal steady state response of the circuit it is necessary to perform algebraic operation such as addition, subtraction, multiplication and division of two or more sinusoidal quantities of the same frequency, different magnitude and phase. Consider two sinusoidal currents whose equations are:

$i_1 = I_{m1} \sin\omega t$

$i_2 = I_{m2} \sin(\omega t+\theta_2)$

The current i_2 leads i_1 by the relative phase angle θ_2. The resultant current i_3 due to addition can be written as

$i_3 = I_{m1} \sin\omega t + I_{m2} \sin(\omega t + \theta_2)$

$= I_{m3} \sin(\omega t + \theta_3)$

This addition results into another sinusoid which can be expressed in terms of a resultant amplitude and phase. The angle θ_3 is the phase angle measured with respect to the same reference point used for θ_2.

This analytical procedure becomes more cumbersome if there are more than two sinusoidal quantities. Again the resultant equation does not present the picture of what is happening in the circuit.

In 1983, Charles P. Steinmetz presented the idea of representing sinusoidal by using a constant amplitude line whose length is equal to maximum value and rotates in ***anticlockwise*** direction at a angular velocity, ω. Such a rotating line is called a ***phasor***. The horizontal phasor represents zero of the sinusoidal waveform and as it rotates in anticlockwise direction a sinusoidal waveform is drawn as per the equation $i = I_{max} \sin\omega t$. The vertical length represents the value of sinusoidal quantity at that instant. The phasor addition or subtraction quickly gives the numerical value and leading or lagging angle.

The phasors are added in the same manner as forces are added. The phasors of same kind are added, i.e., voltage to voltage or current to current. The addition or subtraction can be accomplished by the ***parallelogram*** method.

3.4 In Phasor

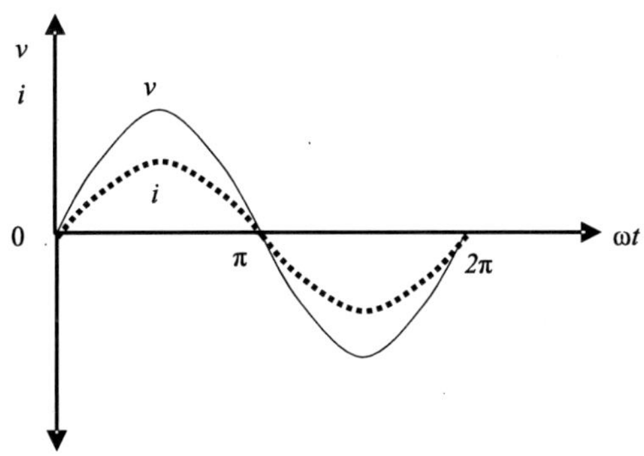

(a) Waveform diagram

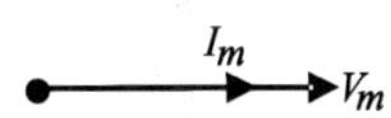

(b) Phasor diagram

Fig. 19.2. Phase difference

The two quantities are said to be in phase with each other if they pass through zero values at the same instant and rise in the same direction as shown in Fig. 19.2. The angle between two phasors is zero degree.

3.5 In Phase Opposition

The two quantities are said to be in phase opposition with each other if they pass through zero values at the same instant and rise in the opposite direction as shown in Fig. 19.3. The phase difference is 180°. When they have a phase difference of 90° or radians, then they are said to be in quadrature.

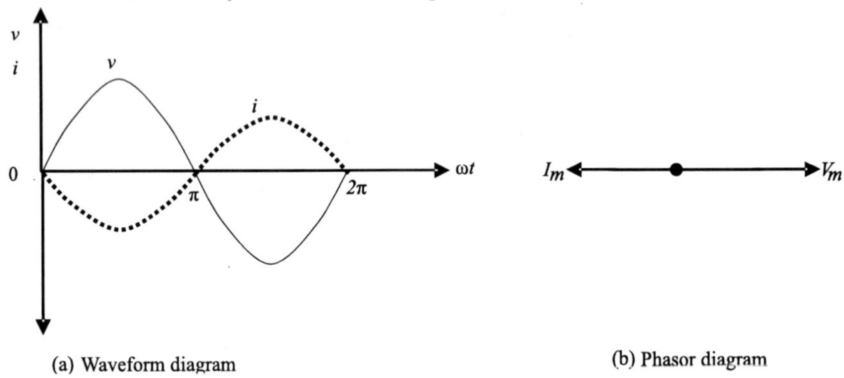

Fig. 19.3. Phase opposition

3.6 Conventions

In order to achieve consistent and accurate results it is essential to follow certain conventions, Such as

1. Counterclockwise direction of rotation of phasor is taken as positive direction of rotation of phasors, i.e., a phasor rotated in a ***counter-clockwise*** direction from a given phasor is said to ***lead*** the given phasor while a phasor rotated in a ***clockwise direction*** is said to ***lag*** the given phasor.
2. For ***series*** circuit, in which the current is common to all parts of the circuit, the current phasor is usually taken as ***reference phasor*** for other phasors in the same diagram and is drawn on horizontal line.
3. For ***parallel*** circuit, in which the voltage is common to all parts of the circuit, the voltage phasor is usually taken as ***reference phasor*** for other phasors in the same diagram and is drawn on horizontal line.
4. It is not necessary that voltage and current phasors are drawn to the same scale. But in one diagram all voltage phasors should be drawn to the same scale.

3.7 Sinusoidal Steady State Response of Single Elements

3.7.1 AC Circuit Containing Resistance Only

The closed path followed by alternating current is called AC circuit. A purely resistive circuit is shown in Fig.19.4(a). Ordinary filament lamps, water resistances etc, are the examples of resistive loads. The whole of the applied voltage is utilized in overcoming the ohmic resistance of the circuit. An alternating or sinusoidal voltage is represented by

$v = V_m \sin\omega t$ or $v = I_{mag}[V_m e^{j\omega t}]$

An alternating or sinusoidal current flowing through a resistive circuit is given by

$i = I_m \sin(\omega t + \theta)$ or $i = I_{mag}[I_m e^{j\omega t} e^{j\theta}]$

where I_m and θ are the amplitude and phase of the response.

Now as per KVL, $V = R\,i$

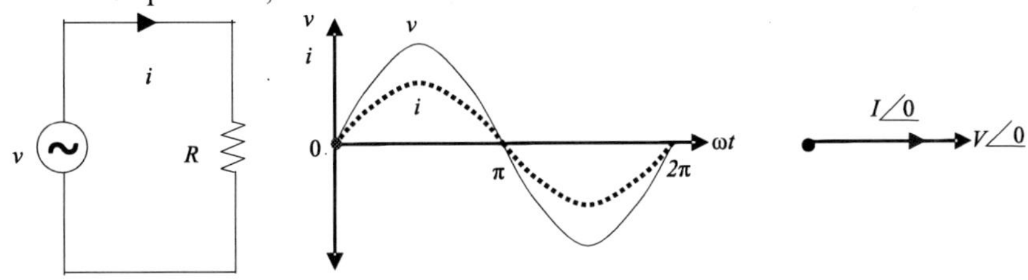

(a) Circuit diagram (b) Waveform diagram (c) Phasor diagram

Fig. 19.4. Resistive circuit connected to AC source

$I_{mag}[V_m e^{j\omega t}] = R\, I_{mag}[I_m e^{j\omega t} e^{j\theta}]$

$V_m e^{j\omega t} = R\, I_m e^{j\omega t} e^{j\theta}$

$I_m e^{j\theta} = [V_m / R] = [V_m / R]\, e^{j0}$

Therefore, $\theta = 0$.

Thus, on RHS there is only real number. This is a characteristic of a purely resistive circuit. The instantaneous current is given by

$i = [V_m / R] \sin\omega t = I_{max} \sin\omega t$

This implies that for a resistive circuit, the current and voltage are in time phase, i.e., zero angle between the voltage and current. The peak values and zero values occur at the same time instants. This is shown in Fig.19.4(b). The phasor diagram is shown in Fig.19.4(c).

3.7.2 AC Circuit Containing Pure Inductor Only

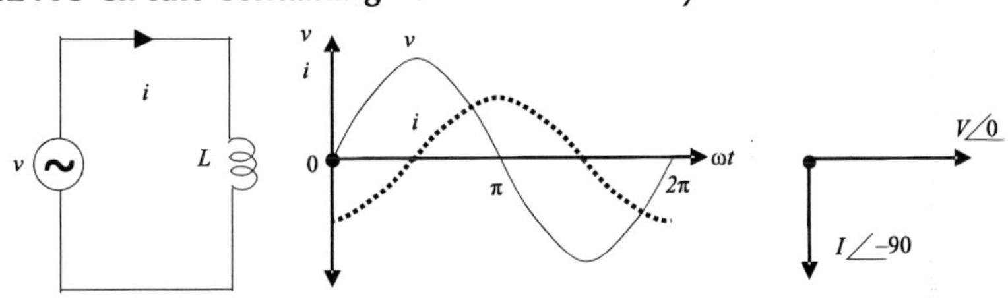

(a) Circuit diagram (b) Waveform diagram (c) Phasor diagram

Fig. 19.5. Pure inductive circuit connected to AC source

Similar to resistor, the potential difference appearing across the inductor terminals is

$v = V_m \sin\omega t$ or $v = I_{mag}[V_m e^{j\omega t}]$

Assuming linear circuit, the response can be represented by

$i = I_m \sin(\omega t + \theta)$ or $i = I_{mag}[I_m e^{j\omega t} e^{j\theta}]$

For the inductive circuit, the voltage equation is

$v = L\, di/dt$

$$I_{mag}[V_m e^{j\omega t}] = L \frac{dI_{mag}[I_m e^{j\omega t} e^{j\theta}]}{dt}$$

$$= L\, I_{mag}[I_m j\omega\, e^{j\omega t} e^{j\theta}]$$

Again for simplicity drop imaginary, we get

$V_m = L\, I_m j\omega\, e^{j\theta}$

$I_m e^{j\theta} = V_m / j\omega L$

The j operator is a rotational operator and is defined as $j = 1 \angle 90$

Any real number which is multiplied by j changes its position from the horizontal axis to the vertical axis.

$$\therefore I_m e^{j\theta} = \frac{V_m}{\Re L \quad 90} = \frac{V_m}{L} \angle \text{ü}$$

$$\therefore I_M = \frac{V_M}{\omega L} \; and \; \theta = -90$$

$$= \frac{V_m}{X_L} \; and \; X_L = \omega L = 2\pi f L \Omega$$

where L is in henry. X_L is called the ***inductive reactance*** of the coil. The term reactance is used to distinguish it from resistance. Inductance limits the magnitude of current in the circuit and the opposition offered by inductance to current flow is X_L. An inductive element causes i to lag behind v by 90° since θ is equal to –90°. The waveform diagram is shown in Fig.19.5(c) and the phasor diagram is shown in Fig.19.5(c).

One example of inductor is choke coil of tubelight. An emf is induced in the coil when AC voltage is applied to the coil. So every time the applied voltage has to overcome this emf. This self-induced emf in the coil is given by

$e_L = -L\, di/dt = -v = -V_m \sin\omega t$

$$\therefore di = \frac{V_m}{L} \sin \omega t\, dt$$

$$\therefore i = \frac{V_m}{L} \int \sin \omega t\, dt = \frac{V_m}{\omega L}(-\cos \omega t)$$

$$i = \frac{V_m}{\omega L} \sin(\omega t - 90) = \frac{V_m}{X_L} \sin(\omega t - \pi/2)$$

–90° or π/2 indicates lagging current by 90° electrical.

3.7.3 AC Circuit Containing Pure Capacitor Only

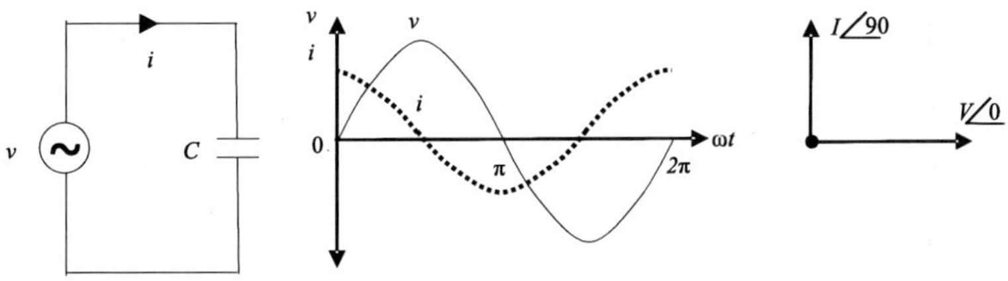

(a) Circuit diagram (b) Waveform diagram (c) Phasor diagram

Fig. 19.6. Pure capacitive circuit connected to AC source

When the DC voltage is impressed across the plates of perfect condenser, it becomes charged to full voltage almost instantaneously. The charging current will flow only during the period of build-up and will cease to flow as soon as the capacitor attains the steady voltage. This shows that for DC, ***capacitor*** acts a ***break*** in the circuit or has infinitely high resistance.

When an alternating voltage is applied across the capacitor, it gets charged in one direction and then in other as the voltage reverses, it is discharged. The result is that electrons move to and fro around the circuit, connecting the plates, thus there is flow of current. The current flowing through the capacitor is given by

$i = C\,dv/dt$ and $v = V_m \sin\omega t$

The current response can be written as

$i = I_m \sin(\omega t + \theta)$

Using exponential form, we get

$$I_m e^{j\theta} e^{j\omega t} = C\frac{d[V_m e^{j\omega t}]}{dt} = CV_m\, j\omega e^{j\omega t}$$

$$I_m e^{j\theta t} = \text{j}\omega CV_m$$

$$I_m \angle\theta = \frac{V_m}{1/\omega C}\angle{+90^\circ}$$

$$\therefore I_m = \frac{V_m}{1/\omega C} \; and \; \theta = +90$$

$$= \frac{V_m}{X_c} \; and \; X_c = 1/\omega C = 1/2\pi f C\,\Omega$$

where C is in farad and X_C is called ***capacitive reactance***. This capacitance not only causes the current to lead the voltage by 90° but it also limits the magnitude of current in the circuit. Capacitance opposes the change in voltage and serves to delay he increase or decrease of voltage across the capacitor. This causes the voltage to lag behind the current.

Applied voltage, $v = V_m \sin\omega t$

Charge on capacitor, $q = Cv = C\,V_m \sin\omega t$

Current, $i = dq/dt = \omega C\,V_m \cos\omega t$

$= \omega C V_m \sin(\omega t + \pi/2)$
$= I_m \sin(\omega t + \pi/2) = I_m \sin(\omega t + 90)$

This indicates that the current leads the voltage by $\pi/2$ or 90°. The waveform diagram is shown in Fig.19.6(b) and the phasor diagram is shown in Fig.19.6(c).

3.7.4 R–L–C Series Circuit

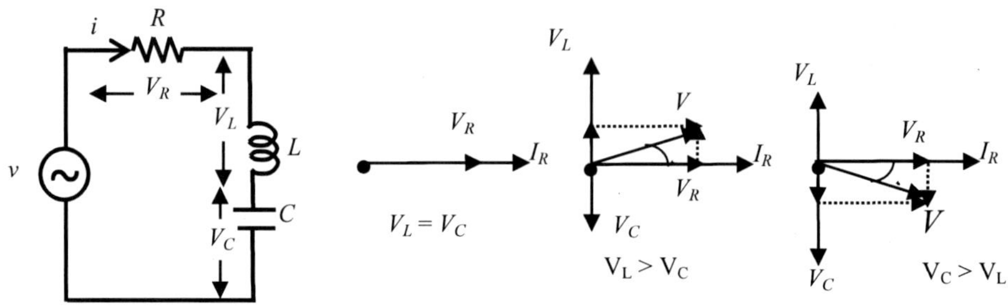

(a) Circuit diagram (b) Phasor diagrams

Fig. 19.7. R-L-C Series circuit connected to AC source

KVL in terms of effective values

$$\bar{V} = \bar{I}R + \bar{I}(j\omega L) + \bar{I}(1/j\omega C) = \bar{I}(R + j\omega L + 1/j\omega C)$$

Complex impedance

$$Z = R + j(\omega L - 1/\omega C) = \sqrt{R^2 + (\omega L - 1/\omega C)^2}\ \angle \tan^{-1}\frac{\omega L - 1/\omega C}{R}$$

The phasor diagram is drawn by taking current as the reference phasor. The voltage across resistance R is V_R, voltage across inductor is V_L and voltage across capacitor is V_C. V_R is in phase with I_R, V_L leads I_R and V_C lags I_R. There are ***three cases*** of R-L-C series circuit:

1. If $X_L = X_C$, $V_L = V_C$, then the circuit is purely resistive. The current and voltage are in phase, $\theta = 0$.
2. If $X_L > X_C$, $V_L > V_C$, then the circuit is inductive. The current lags the applied voltage by an angle θ.
3. If $X_C > X_L$, $V_C > V_L$, then the circuit is capacitive. The current leads the applied voltage by an angle θ.

From the phasor diagram, the applied voltage V can be obtained as

$$V = \sqrt{V_R^2 + (V_L - V_C)^2} = \sqrt{(IR)^2 + (IX_L - IX_C)^2} = I\sqrt{(R)^2 + (X_L - X_C)^2}$$

$$\therefore I = \frac{V}{\sqrt{(R)^2 + (X_L - X_C)^2}} = \frac{V}{Z}$$

The quantity Z offers opposition to current flow and is called ***impedance*** of the circuit.

If the equation for the applied voltage is $v = V_m \sin\omega t$, then equation for the current will be $i = I_m \sin(\omega t \mp \theta)$, $-\theta$ when $X_L > X_C$ and $+\theta$ when $X_C > X_L I$.

3.8 Power Factor

Refer to Fig.19.1. Let $v = V_m \sin\omega t$, and $i = I_m \sin(\omega t - \theta)$, the power P is equal to vi. Therefore

$$P = V_m I_m \sin\omega t \sin(\omega t - \theta)$$

$$= \frac{V_m I_m}{2}\cos\theta - \frac{V_m I_m}{2}\cos(2\omega t - \theta)$$

The average power over one cycle is

$$P_{av} = \frac{V_m I_m}{2}\cos\theta = V_{rms}\, I_{rms}\cos\theta = VI\cos\theta$$

The quantity P is the average power and is expressed in watts. Its value depends on θ in spite of large value of V and I. For this reason the product ***VI*** is called ***apparent power.*** This power is not always realizable in the circuit for doing work. The ***useful part*** depends on the value of cosθ and because of this, it is called the ***power factor*** (***pf***) of the circuit. Thus,

$$Pf = \text{ü} \quad \theta = \frac{\text{Average Power}}{\text{Apparent Power}} = \frac{P}{VI}$$

For R-L-C series circuit the power factor is equal to R/Z.

Power factor of a circuit is a measure of its ***effectiveness in utilizing*** the apparent power drawn by it.

The greater the power factor of a circuit, the greater is its ability to utilize the apparent power. For this reason, it may also be called ***utilization factor*** of apparent power.

If the ***current lags*** to the voltage, the ***power factor is lagging***. If the current leads to the voltage, the power factor is leading. If current and voltage are in phase then the power factor is unity.

VI* cos** θ is called ***active power and is expressed in watts or kW. This is also called ***true power.*** This is the power which is actually consumed in the circuit.

This is consumed in resistance since neither pure inductor nor pure capacitor consumes any active power.

The product of apparent power, ***VI*** and the **sin**θ is called the ***reactive power, Q***. This is also called ***wattless power*** and is expressed in reactive volt-amperes or kilo-volt ampere reactive (kVAR). This is a circulating power. In half cycle this power is consumed by L or C and in the other half cycle it is returned to the supply. This is the power that flows back and forth in both directions in the circuit or ***reacts upon itself***. Hence, the name is given ***reactive***.

$$\therefore \text{Apparent Power} = \sqrt{(\text{Active Power})^2 + (\text{Reactive Power})^2}$$

$$\therefore VI = \sqrt{(VI\,Cos\,\theta)^2 + (VI\,\sin\,\theta)^2} = \sqrt{(kW)^2 + (kVAR)^2}$$

$$S = \sqrt{P^2 + Q^2}$$

4. APPARATUS REQUIRED

1. Single Phase Variac / Dimmerstat / Autotransformer: 230V, 2-4 Amp.
2. AC Ammeter (Moving Iron Type: used for AC and DC): One, 1-2 Amp.
3. AC Voltmeter (Moving Iron Type: used for AC and DC): One, 300V or Multimeter.
4. Wattmeter: One, 2 Amp, 230V.
5. Rheostat: One, 2 Amp rating.
6. Inductor (Variable): One, 2 Amp rating.
7. Capacitor: One or Two, 2 Amp rating.

5. CIRCUIT DIAGRAM

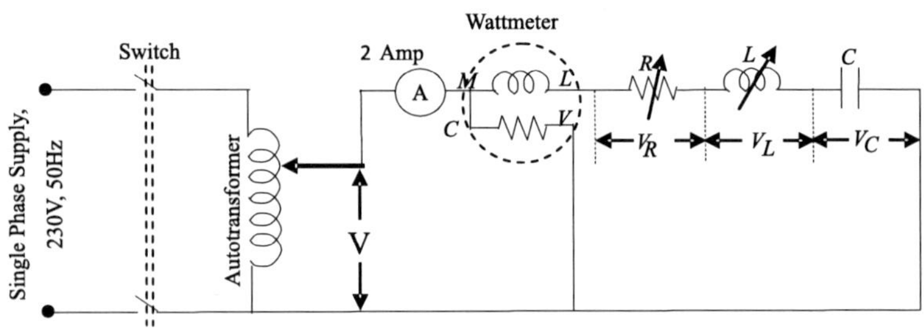

Fig. 19.8. Experimental set-up for studying R-L-C series circuit

6. PROCEDURE

1. Do the connections as per the circuit diagram.
2. After the connections are checked, set autotransformer to zero output voltage and then make an AC supply.
3. Increase output voltage to a suitable value.
4. Note down voltages across autotransformer, *R*, *L* and *C*, current and power.
5. Increase the voltage and repeat step 4.
6. Vary the resistance *R* and inductance *L* such that $V_L = V_C$, $V_L > V_C$ and $V_L < V_C$. Repeat step 4 for all these three conditions.

7. OBSERVATIONS

Multiplying factor of Wattmeter, *M*

$$= \frac{\text{Voltage Rating of Pressure Coil} \times \text{Current Rating of Current Coil of Wattmeter}}{\text{Full Scale Deflection of Wattmter}}$$

Reading of wattmeter, $P = M \times$ deflection of wattmeter.

Sr. No.	V Volt	I Amp	P Watt	V_R Volt	V_L Volt	V_C Volt
1.						
2.						
3.						
4.						

8. CALCULATIONS

Draw the phasor diagrams for all sets of readings.

Sr. No.	$R=V_R/I$ Ohm	$X_L=V_L/I$ Ohm	$X_C=V_C/I$ Ohm	$\text{Cos}\theta = P/VI$	θ calculated	θ from phasor diagram
1.						
2.						
3.						
4.						

9. RESULTS

The relationship for R-L-C series circuit has been verified. The error in calculated θ and θ from phasor diagram is due to error in meters and also due to internal resistance of L and C.

10. CONCLUSION

1. For, $X_L = X_C$, the circuit is ***resistive*** and the power factor is **unity**.
2. For $X_L > X_C$, the circuit is ***inductive*** and the power factor is **lagging**.
3. Greater the value of X_L-X_C/R, greater the value of θ.
4. For $X_C > X_L$, the circuit is ***capacitive*** and the power factor is **leading**.
5. Greater the value of X_C - X_L/R, greater the value of θ.
6. Power is consumed by resistance and not by inductance and capacitance.

11. POST-EXPERIMENTAL QUIZ

1. The voltage, $v = V_m \sin\theta$, write down the equation for current if power factor is lagging and leading?
2. How to draw the phasor diagram for series circuit?
3. Explain apparent power, active power and reactive power?

CHAPTER 20

To Study R-L-C Parallel Circuit

1. OBJECTIVE

The main objective of this experiment is to study the effect of resistive, inductive and capacitive circuit on voltage, current phasor and power factor.

2. PRE-EXPERIMENTAL QUIZ

1. Define complex power?
2. What is the practical significance of apparent power?
3. How will you convert the parallel circuit to series circuit?

3. THEORY

Refer theory of R-L-C series circuit.

The closed path followed by alternating current is called an AC circuit. When the numbers of electrical components or elements are connected in such a way that one end of each of them is joined to a common point and the other ends being joined to another common point, then the circuit is called **parallel circuit**. In this circuit the current is divided into as many as parallel paths.

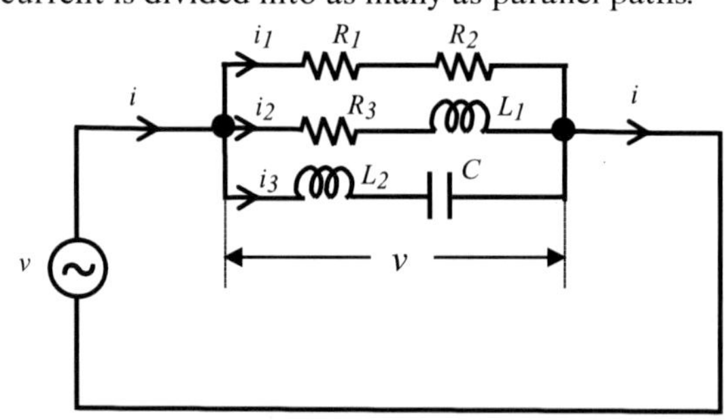

Fig. 20.1. Parallel circuit

The most useful property of a parallel circuit is the fact that potential difference has the same value between the terminals of each branch of parallel circuit. Parallel circuits are very common in use. Generally, all loads are connected in parallel. The series circuit is an "ALL or NONE" circuit in which either everything operates or nothing operates. The parallel circuit offers the following advantages:

1. The appliances rated for same voltage but different powers can be connected in parallel ***without disturbing each other's performance***. Thus, a 230 V, 200 W TV receiver can be operated independently in parallel with a 230 V 40/60 W lamp.

2. If a break occurs in any one of the branch circuits, it will have no effect on the other branch circuits.
3. ***Individual*** circuit or load control is possible, independently.

Some points may be noted:

1. Same voltage acts across all parts of the circuit.
2. The parallel branches have their individual currents.
3. The total current is equal to the ***phasor*** sum of individual currents through the various branches.
4. Powers are additive.

3.1 Phasor Representation of Sinusoidal Quantities

Often in determining the sinusoidal steady state response of the circuit it is necessary to perform algebraic operation such as addition, subtraction, multiplication and division of two or more sinusoidal quantities of the same frequency, different magnitude and phase.

In 1983, Charles P. Steinmetz had presented the idea of representing sinusoidal by using a constant amplitude line whose length is equal to maximum value and rotates in ***anticlockwise*** direction at a angular velocity, ω. Such a rotating line is called ***phasor***. The horizontal phasor represents zero of the sinusoidal waveform and as it rotates in anticlockwise direction a sinusoidal waveform is drawn as per the equation $i = I_{max} \sin \omega t$. The vertical length represents the value or magnitude of sinusoidal quantity at that instant. The phasor addition or subtraction quickly gives the numerical value and leading or lagging angle.

The phasors are added in the same manner as forces are added. The phasors of same kind are added, i.e., voltage to voltage or current to current. The addition or subtraction can be accomplished by the ***parallelogram*** method.

3.2 Conventions

In order to achieve consistent and accurate results it is essential to follow certain conventions, such as

1. ***Counterclockwise direction*** of rotation of phasor is taken as ***positive*** direction of rotation of phasors, i.e., a phasor rotated in a ***counter-clockwise*** direction from a given phasor is said to ***lead*** the given phasor while a phasor rotated in a ***clockwise direction*** is said to ***lag*** the given phasor.
2. For ***series*** circuit, in which the current is common to all parts of the circuit, the current phasor is usually taken as ***reference phasor*** for other phasors in the same diagram and is drawn on horizontal line.
3. For ***parallel*** circuit, in which the voltage is common to all parts of the circuit, the voltage phasor is usually taken as ***reference phasor*** for other phasors in the same diagram and is drawn on horizontal line.
4. Each ***branch*** of parallel circuit is considered a ***series circuit*** and the current phasor is then considered a reference phasor for that branch only. The effect of all branches can be combined together.

5. The ***length of phasor*** is equal to the ***maximum value*** of alternating quantity but sometimes it can be taken as equal to **rms** value since the voltmeter and ammeter measure rms value.
6. It is not necessary that voltage and current phasors are drawn to the same scale. But in one diagram all voltage phasors should be drawn to the same scale.

3.3 Sinusoidal Steady State Response of Single Elements

3.3.1 AC Circuit Containing Resistance Only

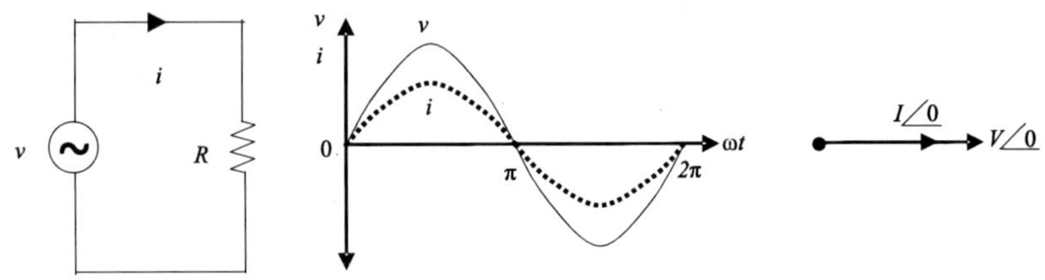

(a) Circuit diagram (b) Waveform diagram (c) Phasor diagram

Fig. 20.2. Resistive circuit connected to AC source

A purely resistive circuit, its waveform diagram and phasor diagram are shown in Fig. 20.2. Ordinary filament lamps, water resistances, etc. are the examples of resistive loads. The whole of the applied voltage is utilized in overcoming the ohmic resistance of the circuit. An alternating or sinusoidal voltage is represented by

$v = V_m \sin\omega t$ or $v = I_{mag}[V_m e^{j\omega t}]$

An alternating or sinusoidal current flows through the resistive circuit and it is given by

$i = I_m \sin(\omega t + \theta)$ or $i = I_{mag}[I_m e^{j\omega t} e^{j\theta}]$

where I_m and θ are the amplitude and phase of the response.

Now as per KVL, $V = Ri$

$$I_{mag}[V_m e^{j\omega t}] = R\, I_{mag}[I_m e^{j\omega t} e^{j\theta}]$$

$$V_m e^{j\omega t} = R\, I_m e^{j\omega t} e^{j\theta}$$

$$I_m e^{j\theta} = [V_m / R] = [V_m / R]\, e^{j0}$$

Therefore, $\theta = 0$.

Thus on RHS there is only real number. This is a characteristic of purely resistive circuit. The instantaneous current is given by

$$i = [V_m / R] \sin\omega t = I_{max} \sin\omega t$$

This implies that for a resistive circuit, the current and voltage are in time phase, i.e., zero angle between voltage and current. The peak values and zero values occur at the same time instants.

3.3.2 AC Circuit Containing Pure Inductor Only

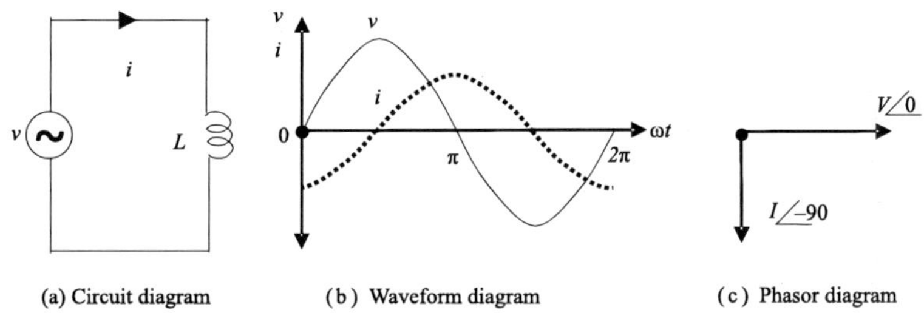

Fig. 20.3. Pure inductive circuit connected to AC source

Similar to resistor, the potential difference appearing across the inductor terminals is

$v = V_m \sin\omega t$ or $v = I_{mag}[V_m e^{j\omega t}]$

Assuming linear circuit, the response can be represented by

$i = I_m \sin(\omega t+\theta)$ or $i = I_{mag}[I_m e^{j\omega t} e^{j\theta}]$

For the inductive circuit, the voltage equation is

$v = L\, di/dt$

$$I_{mag}[V_m e^{j\omega t}] = L \frac{dI_{mag}[I_m e^{j\omega t} e^{j\theta}]}{dt}$$

$$= L\, I_{mag}[I_m j\omega\, e^{j\omega t} e^{j\theta}]$$

Again for simplicity drop imaginary, we get

$V_m = L\, I_m j\omega\, e^{j\theta}$

$I_m\, e^{j\theta} = V_m / j\omega L$

The j operator is a rotational operator and is defined as $j = 1 \angle 90$.

Any real number which is multiplied by j changes its position from the horizontal axis to the vertical axis.

where L is in henry. X_L is called the ***inductive reactance*** of the coil. The term reactance is used to distinguish it from resistance. Inductance limits the magnitude of current in the circuit and the opposition offered by inductance to current flow is X_L. An inductive element causes i to lag behind v by 90° since θ is equal to –90°.

The waveform diagram is shown in Fig. 20.3(c) and the phasor diagram is shown in Fig. 20.3(c).

The example of inductor is choke coil of tubelight. An emf in induced in the coil when AC voltage is applied to the coil. So every time the applied voltage has overcome this emf.

This self-induced emf in the coil is given by

$$e_L = -L\, di / dt = -v = V_m \sin \omega t$$

$$\therefore di = \frac{V_m}{L} \sin \omega t\ dt$$

$$\therefore i = \frac{V_m}{L} \int \sin \omega t\ dt = \frac{V_m}{\omega L}$$

$$i = \frac{V_m}{\omega L} \sin (\omega t - 90) = \frac{V_m}{X_L} \sin (\omega t - \pi / 2)$$

–90° or π/2 indicates lagging current by 90° electrical.

3.3.3 AC Circuit Containing Pure Capacitor Only

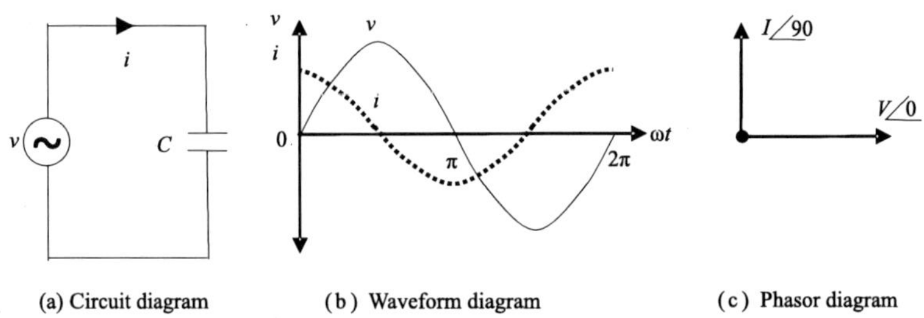

Fig. 20.4. Pure Capacitive Circuit Connected to AC Source

When the DC voltage is impressed across the plates of perfect condenser, it becomes charged to full voltage almost instantaneously. The charging current flows only during the period of build-up and ceases to flow as soon as the capacitor attains the steady voltage. This shows that for DC, ***capacitor*** acts a ***break*** in the circuit or has infinitely high resistance.

When an alternating voltage is applied across the capacitor, it gets charged in one direction and then in other as the voltage reverses, it is discharged. The result is that electrons move to and fro around the circuit, connecting the plates, thus there is flow of current.

The current flowing through the capacitor is given by

$i = C\, dv/dt$ and $v = V_m \sin\omega t$

The current response can be written as

$i = I_m \sin(\omega t + \theta)$

Using exponential form, we get

$$I_m e^{j\theta} e^{j\omega t} = C \frac{d[V_m e^{j\omega t}]}{dt} = CV_m\ j\omega\ e^{j\omega t}$$

$$I_m e^{j\theta t} = j\,\omega\, C V_m$$

$$I_m \angle\theta = \frac{V_m}{1/\omega C} \angle + 90°$$

$$\therefore I_m = \frac{V_m}{1/\omega C} \text{ and } \theta = +90$$

$$I_m = \frac{V_m}{X_C} \text{ and } X_C = 1/\omega C = 1/2\pi f c\Omega$$

where C is in farad and X_C is called **capacitive reactance**. This capacitance not only causes the current to lead the voltage by 90° but it also limits the magnitude of current in the circuit. Capacitance opposes the change in voltage and serves to delay the increase or decrease of voltage across the capacitor. This causes the voltage to lag behind the current.

Applied voltage, $v = V_m \sin\omega t$

Charge on capacitor, $q = Cv = CV_m \sin\omega t$

Current, $i = dq/dt = \omega C\, V_m \cos\omega t$

$= \omega C\, V_m \sin(\omega t+\pi/2)$

$= I_m \sin(\omega t+\pi/2) = I_m \sin(\omega t + 90)$

This indicates that the current leads the voltage by $\pi/2$ or 90°.

The waveform diagram is shown in Fig. 20.4(c) and the phasor diagram is shown in Fig.20.4(c).

3.3.4 R–L Series Circuit

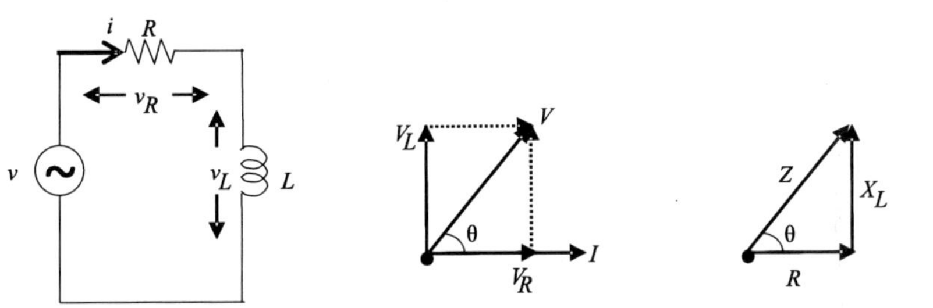

(a) Circuit diagram (b) Phasor diagram (c) Impedance triangle

Fig. 20.5. R-L-C series circuit connected to AC source

Figure 20.5 (a) shows a pure resistance, R ohms connected in series with a coil of pure inductance, L henry. This is most common inductive circuit used in application. 80% of the load is having this type of circuit.

For this circuit

$$v = Ri + L\frac{di}{dt}$$

$$V_m \sin \omega t = RI_m \sin(\omega t - \theta) + L\frac{dI_m \sin(\omega t - \theta)}{dt}$$

$$V_m e^{j\omega t} = RI_m e^{-j\theta} e^{j\omega t} + I_m Lj\omega e^{-j\theta} e^{j\omega t}$$

$$V_m = I_m e^{-j\theta} (R + j\omega L)$$

$$\vec{V} = \vec{I} e^{-j\theta} (R + jX_L)$$

If V and I are the rms values, then $V_R = IR$ and $V_L = IX_L$, where V_R is in phase with I and V_L leads I by 90°. Taking common current I as the reference phasor, the phasor diagram of the circuit is drawn and is shown in Fig. 20.5 (b).

The applied voltage V is the phasor sum of V_R and V_L.

$$V = \sqrt{V_R^2 + V_L^2} = \sqrt{(IR)^2 + (IX_L)^2} = I\sqrt{R^2 + X_L^2} = IZ_L \text{ or Simply } IZ$$

$$Z = \sqrt{R^2 + X_L^2} = R + j\omega L$$

The quantity Z offers opposition to current flow and is called ***impedance*** of the circuit. It is clear from the phasor diagram that the circuit current ***I lags*** behind the voltage V by an angle θ. The value of this phase angle θ can be determined from the phasor diagram.

$\theta = \tan^{-1}\dfrac{X_L}{R}$ and the power factor of the circuit is cosθ. From the impedance triangle, as shown in Fig. 20.5 (c), the power factor, $\cos\theta = R / Z$. The greater the value of (X_L / R), greater will be the value of phase angle θ. The value of phase angle θ will be more than 0° and less than 90°.

3.3.5 R-C Series Circuit

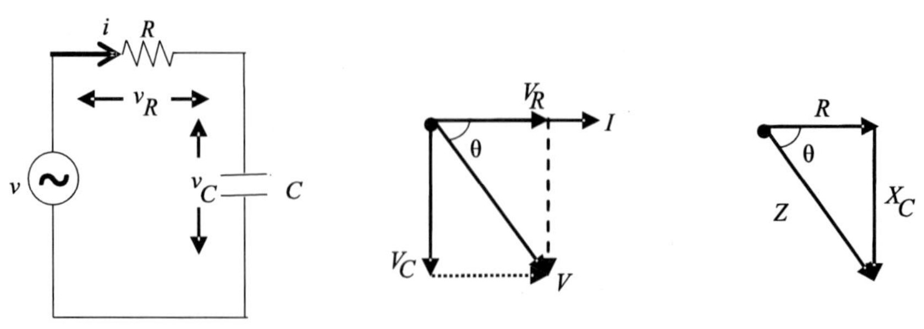

(a) Circuit diagram (b) Phasor diagram (c) Impedance triangle

Fig. 20.6. R - C Series Circuit Connected to AC Source

Figure 20.6 (a) shows a pure resistance, R ohms connected in series with a capacitor, C farad.

For this circuit

$$v = Ri + \frac{1}{C}\int i\, dt$$

$$V.\ \sin \omega t = RI \quad \sin(\omega t + \theta) + \frac{1}{C}\int I \quad \sin(\omega t + \theta)\, dt$$

$$V.\ e^{j\omega t} = RI \quad e^{j\theta}\, e^{j\omega t} + I \quad \frac{1}{C\, j\omega} e^{j\theta}\, e^{j\omega t}$$

$$V_m \ \Re I_m\, e^{j\theta}\,(R \quad \frac{1}{Cj\omega}) \quad I_m\, e^{j\theta}\,(R - jX_C)$$

ð

$$V = I\, e^{j\theta}\ \text{ü}R \quad jX_C$$

If V and I are the rms values, then $V_R = IR$ and $V_C = IX_C$, where V_R is in phase with I and V_C lags I by 90°. Taking common current I as the reference phasor, the phasor diagram of the circuit is drawn and is shown in Fig. 20.6 (b).

The applied voltage V is the phasor sum of V_R and V_C.

$$V = \sqrt{V_R^2 + V_C^2} = \sqrt{(IR)^2 + (IX_C)^2} = I\sqrt{R^2 + X_C^2} = IZ_C \text{ or simply } IZ$$

$$Z = \sqrt{R^2 + X_C^2} = R - j\frac{1}{\omega C}$$

The quantity Z offers opposition to current flow and is called ***impedance*** of the circuit. It is clear from the phasor diagram that the circuit current ***I leads*** the applied voltage V by an angle θ. The value of this phase angle θ can be determined from the phasor diagram.

$\theta = \tan^{-1}\dfrac{X_C}{R}$ and the power factor of the circuit is $\cos\theta$. From the impedance triangle, as shown in Fig. 20.6 (c), the power factor, $\cos\theta = R / Z$. The greater the value of (X_C / R), greater will be the value of phase angle θ. The value of phase angle θ will be more than 0° and less than 90°.

3.3.6 R-L-C Parallel Circuit

For solving the parallel AC circuits, there are four methods, namely:

1. Phasor method
2. Phasor algebra method or j method
3. Impedance method
4. Admittance method

In the phasor method, the magnitude and phase angle of each branch is calculated. The phasor diagram is drawn by taking common quantity. The phasor addition of voltage or current is obtained either by the parallelogram method or by the component method. Consider the parallel circuit consisting of two branches and connected to an alternating voltage of V volts as shown in Fig. 20.7 (a). The branch 1 consists of rheostat, R and inductor, L. The equivalent impedance Z_{RL} is

$$Z_{RL} = \sqrt{R^2 + X_L^2}$$

The current i_{RL} is

$i_{RL} = V / Z_{RL}$

and the angle θ_{RL} is

$$\theta_{RL} = \tan^{-1}\frac{X_L}{R}$$

The current i_{RL} lags the applied voltage V by an angle θ_{RL} due to inductive circuit. For the branch 1, current i_{RL} is common. So first draw the horizontal line representing the phasor for current I_{RL}. V_R is in phase and V_L leads by 90° to this current I_{RL}. The resultant voltage is V. In parallel circuit the same voltage is applied to capacitor, C. Now draw the current I_C, leading to the voltage V by 90°.

The resultant current, I is the phasor sum of I_{RL} and I_C or it is given by

$$I = \sqrt{I_{RL}^2 + I_C^2 + 2I_{RL}\,I_C\cos(\theta_{RL} + 90)}$$

Depending on the magnitude of currents there are three cases of R-L-C circuit.

1. $I_C = I_{RL}\sin\theta_{RL}$
   The resultant current, I is in phase with the applied voltage V and power factor is unity. The phasor diagram is shown in Fig. 20.7 (b).
2. $I_C < I_{RL}\sin\theta_{RL}$
   The resultant current, I is lagging to the applied voltage V by an angle θ_{PF} and power factor is lagging. The Power factor is $\cos\theta_{PF}$. The phasor diagram is shown in Fig. 20.7 (c).
3. $I_C > I_{RL}\sin\theta_{RL}$
   The resultant current, I is leading to the applied voltage V by an angle θ_{PF} and power factor is leading. The Power factor is $\cos\theta_{PF}$. The phasor diagram is shown in Fig. 20.7 (d).

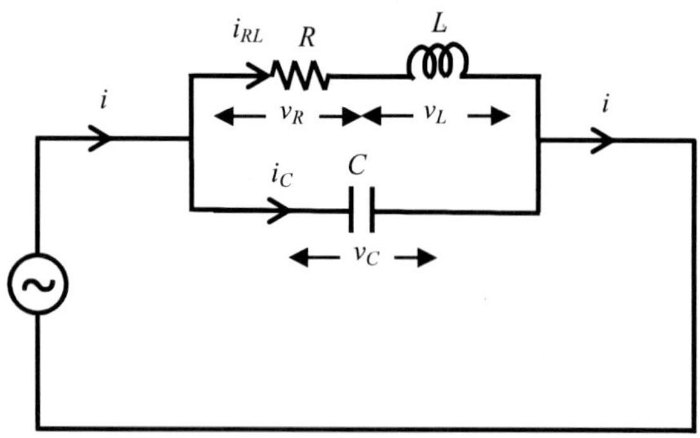

(a) Circuit Diagram

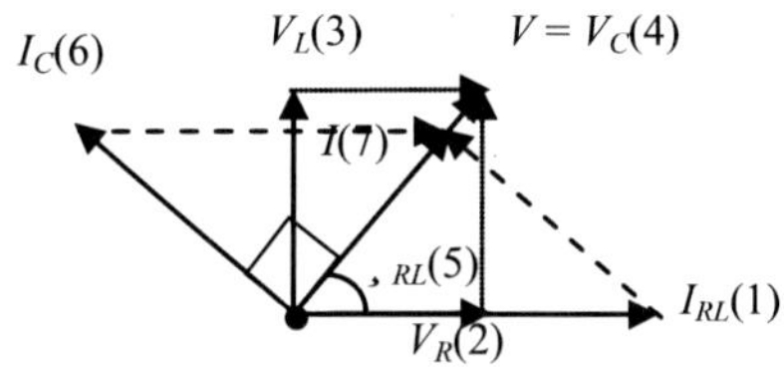

(b) I and V are in Phase, $I_C = I_{RL}sin$, $_{RL}$, Unity Power Factor

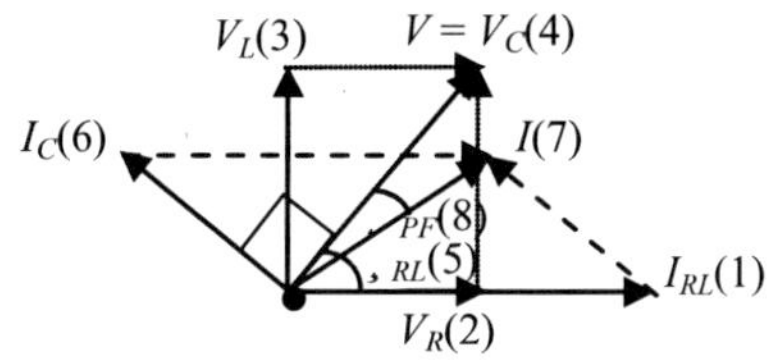

(c) I is lagging to V, $I_C < I_{RL}sin$, $_{RL}$, Lagging Power Factor

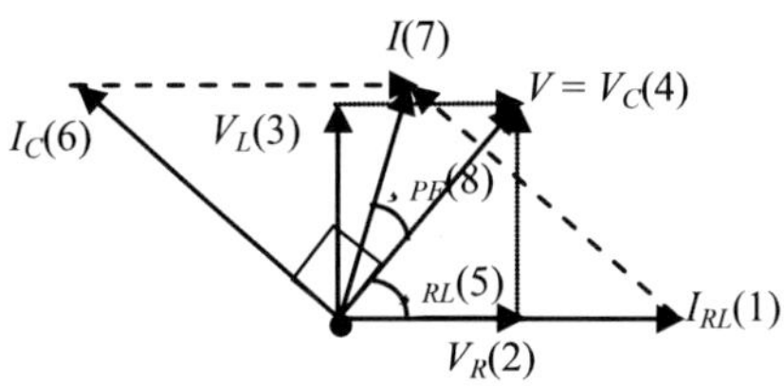

(d) I is leading to V, $I_C > I_{RL}sin$, $_{RL}$, Leading Power Factor

Fig. 20.7. R-L-C Parallel circuit connected to AC source

4. APPARATUS REQUIRED

1. Single Phase Variac / Dimmerstat / Autotransformer: 230 V, 2–4 Amp.
2. AC Ammeter (Moving Iron Type: used for AC and DC): Three, 2-5 Amp.
3. AC Voltmeter (Moving Iron Type: used for AC and DC): One, 300 V or Multimeter.
4. Wattmeter: One, 2–5 Amp, 230 V.
5. Rheostat: One, 2 –5 Amp rating.
6. Inductor (Variable): One, 2–5 Amp rating.
7. Capacitor: One or Two, 2–5 Amp rating.

5. CIRCUIT DIAGRAM

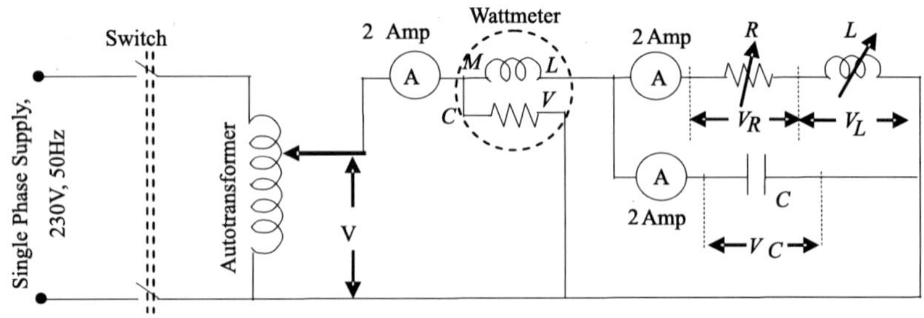

Fig. 20.8. Experimental setup for studying R-L-C parallel circuit

6. PROCEDURE

1. Make the connections as per the circuit diagram.
2. After the connections are checked, set autotransformer to zero output voltage and then make an AC supply.
3. Increase output voltage to a suitable value.
4. Note down voltages across autotransformer, *R*, *L* and *C*, current and power.
5. Increase the voltage and repeat step 4.
6. Vary the resistance *R* and inductance *L* such that $I_C = I_{RL} \sin\theta_{RL}$, $I_C < I_{RL} \sin\theta_{RL}$ and $I_C > I_{RL} \sin\theta_{RL}$. Repeat step 4 for all these three conditions.

7. OBSERVATIONS

Multiplying factor of wattmeter, *M*

$$= \frac{\text{Voltage rating of pressure coil} \times \text{current rating of current coil of wattmeter}}{\text{Full scale deflection of wattmter}}$$

Reading of wattmeter, $P = M$ x deflection of wattmeter.

Sr. No.	V Volt	P Watt	I Amp	I_{RL} Amp	I_C Amp	V_R Volt	V_L Volt	V_C Volt
1.								
2.								
3.								
4.								

8. CALCULATIONS

Draw the phasor diagrams for all sets of readings.

Sr. No.	$R = V_R / I_{RL}$ Ohm	$X_L = V_L / I_{RL}$ Ohm	$X_C = V_C / I_C$ Ohm	$\theta_{RL} = \tan^{-1} X_L/R$	$\cos\theta_{PF} = P/VI$	θ_{PF} calculated	θ_{PF} from phasor diagram
1.							
2.							
3.							
4.							

9. RESULTS

The relationship for R-L-C parallel circuit has been verified. The error in calculated θ_{PF} and θ_{PF} from phasor diagram is due to error in metres and also due to internal resistance of L and C.

10. CONCLUSION

1. For, $I_C = I_{RL} \sin\theta_{RL}$, the circuit is **resistive** and the power factor is **unity**.
2. For $I_C < I_{RL} \sin\theta_{RL}$, the circuit is **inductive** and the power factor is **lagging**.
3. For $I_C > I_{RL} \sin\theta_{RL}$, the circuit is **capacitive** and the power factor is **leading**.
4. Power is consumed by resistance and not by inductance and capacitance.

11. POST-EXPERIMENTAL QUIZ

1. How you will calculate the equivalent impedance of parallel circuit?
2. What are the conventions for drawing the phasor diagram of parallel circuit?
3. How you will calculate the total current in case of parallel circuit?

Bibliography

1. Kothari D.P. and I.J. Nagrath, *Electric Machines*, 4th edition, Tata McGraw-Hill Publishing Company Ltd., New Delhi, 2010.
2. Kothari D.P. and I.J. Nagrath, *Electric Machines*, (Sigma Series) 2nd edition, Tata McGraw-Hill Publishing Company Ltd., New Delhi, 2006.
3. Kothari D.P. and I.J. Nagrath, *Basic Electrical Engineering*, 3rd edition, Tata McGraw-Hill Publishing Company Ltd., New Delhi, 2010.
4. Bimbhra P.S., *Electrical Machinery*, 7th edition, Khanna Publishers, Delhi, 2011.
5. Bimbhra P.S., *Generalized Theory in Electrical Machines*, 5th edition, Khanna Publishers, Delhi, 2011.
6. Fitzerald A.E., Charles Kingsley, Jr. Stephen D. Umans, *Electrical Machinery*, 6th edition, Tata McGraw-Hill Publishing Company Ltd., New Delhi, 2002.
7. Mukharjee P.K. and S. Chakravorti, *Electric Machines*, Dhanpat Rai & Sons, Delhi, 1990.
8. Ashfaq Husain, *Electric Machines*, 2nd edition, Dhanpat Rai & Company, Delhi, 2011.
9. Thereja B.L., *Electrical Technology*, Niraj Prakashan, Delhi, 1970.
10. Tarnekar S.G. and P.K. Kharbanda, *Laboratory Courses in Electrical Engineering*, 2nd edn., S. Chand and Company Ltd., New Delhi, 2003.
11. Murthy S.S., B.P. Singh, C.S. Jha and D.P. Kothari, *Laboratory Manual for Electromechanics*, Wiley Eastern, New Delhi, 1982.

Index